Foundation Fieldbus

2nd Edition

Foundation Fieldbus

2nd Edition

by Ian Verhappen
and Augusto Pereira

ISA™

Notice

The information presented in this publication is for the general education of the reader. Because neither the author nor the publisher have any control over the use of the information by the reader, both the author and the publisher disclaim any and all liability of any kind arising out of such use. The reader is expected to exercise sound professional judgment in using any of the information presented in a particular application.

Additionally, neither the author nor the publisher have investigated or considered the affect of any patents on the ability of the reader to use any of the information in a particular application. The reader is responsible for reviewing any possible patents that may affect any particular use of the information presented.

Any references to commercial products in the work are cited as examples only. Neither the author nor the publisher endorse any referenced commercial product. Any trademarks or tradenames referenced belong to the respective owner of the mark or name. Neither the author nor the publisher make any representation regarding the availability of any referenced commercial product at any time. The manufacturer's instructions on use of any commercial product must be followed at all times, even if in conflict with the information in this publication.

Printed in the United States of America.
10 9 8 7 6 5 4

ISBN 1-55617-964-2

ISA
67 Alexander Drive
P.O. Box 12277
Research Triangle Park, NC 27709
www.isa.org

Library of Congress Cataloging-in-Publication Data
Verhappen, Ian.
Foundation fieldbus / by Ian Verhappen and Augusto Pereira. -- 2nd ed.
p. cm.
Includes bibliographical references and index.
ISBN 1-55617-964-2
1. Intelligent control systems. 2. Process control. 3. Industrial electronics. 4. Computer networks. I. Pereira, Augusto. II. Title.
TJ217.5 .V47 2005
629.8--dc22

2005028785

For Michele, Ashley, & Madeline who watched me create this work and gave me time to do it. Love, "Dad"

Thank you also to everyone who has supported me during my Fieldbus development. You have made it possible to be able to share this knowledge.

Ian Verhappen

For my wife Margareth and my sons Sergio and Fabio that gave me all the support to write this book.

Thanks to everyone who helped me to learn Fieldbus concepts and during all the several projects that I was involved up to now.

Augusto Pereira

Table of Contents

List of Figures

List of Tables

Foreword

This book is a handy reference to FOUNDATION™ Fieldbus that we hope you find useful as you explore and implement this exciting technology for the 21st century. It represents a summary of background information, tips, tricks, and items to note when installing, maintaining, or troubleshooting a FOUNDATION™ Fieldbus system. What this publication is not is a complete "how to" about everything there is to know in this constantly evolving and improving technology.

Since the original publication of this book and its translation into Portuguese and Spanish, available from ISA District 4, we are pleased not only by the first edition's acceptance but also by Fieldbus technology itself. We hope that in some small way the knowledge that we have been able to share here has helped contribute to that success.

As with all things in the automation and controls industry, Fieldbus technology has not been "standing still" so this edition includes updates on some of those changes including Electronic Device Description Language, Fieldbus Safety, High Speed Ethernet (HSE), and additional information on techniques to design networks in different environments like FISCO and FNICO.

More complete coverage of this technology is available through other ISA publications and the standards themselves, which are available from ISA, IEC, and the Fieldbus Foundation.

We hope you enjoy the new larger format of this edition as well. If you have any comments or questions, please contact us so that we can continue to help you as you progress through your Fieldbus journey.

Ian Verhappen
Augusto Periera

1 – Fieldbus Layers

FOUNDATION™ Fieldbus has several different "layers." This chapter discusses three of these layers:

1. Physical Layer: The various topologies and types of data blocks used by FOUNDATION™ Fieldbus
2. Communication Layer: How Fieldbus uses and assigns device registers
3. Parameter Classes: The function or role of the information generated on the network

This chapter provides the background on the how and what Fieldbus is. So let's start. What is Fieldbus?

Fieldbus is a bi-directional digital communication network that enables the connection of multiple field instruments and processes and operator stations (HMI). They carry out control functions and enable monitoring by means of supervision software. Figure 1-1 shows how these three layers (Field, Fieldbus, and Supervisory System) interrelate.

The FOUNDATION™ Fieldbus protocol was based on the ISO/OSI seven-layer communications model, although it does not include all layers. It can be divided into the *Physical Layer* (dealing with instrument connection techniques) and the *Communication Stack* (dealing with the digital communication among the devices). These are the OSI layers used by FOUNDATION™ Fieldbus. Figure 1-2 represents how the FOUNDATION™ Fieldbus protocol maps to the OSI model.

Figure 1-1 – Digital control system architecture.

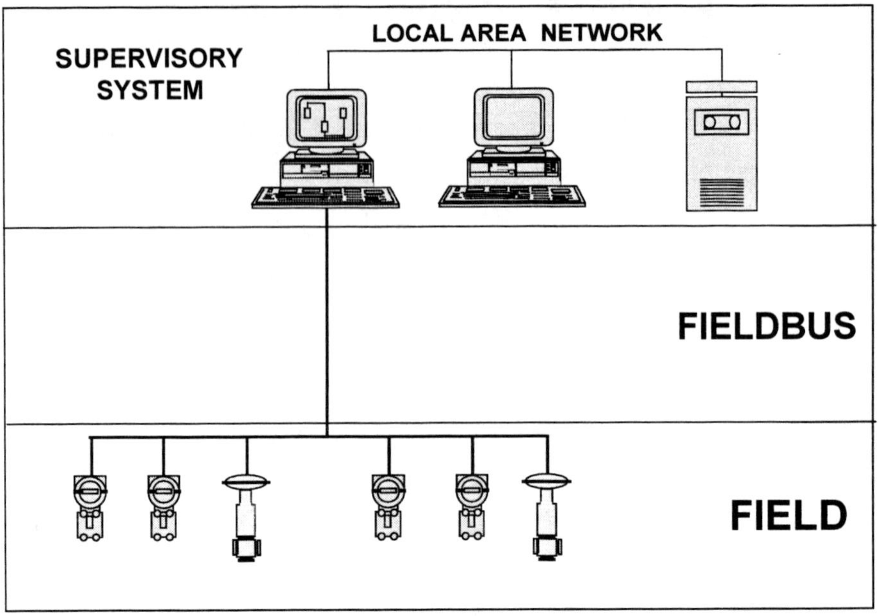

Figure 1-2 – OSI model compared with Fieldbus model.

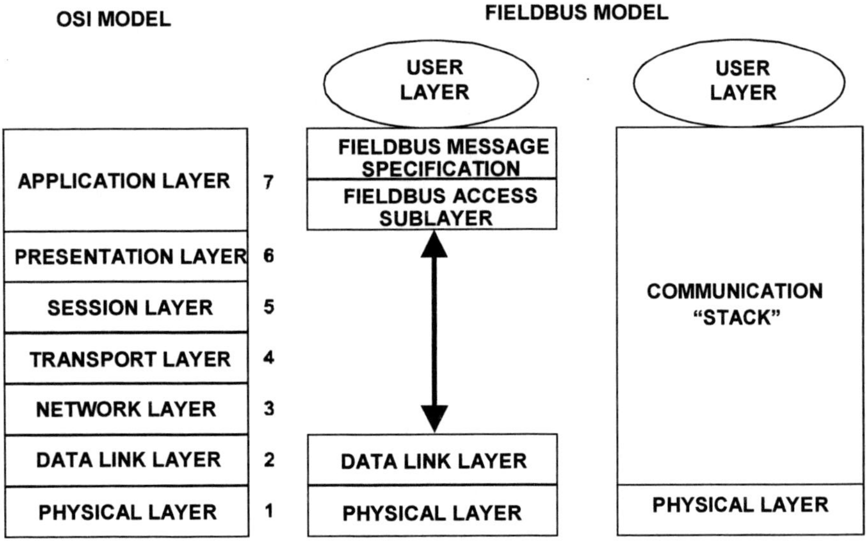

Figure 1-3 represents Manchester Encoding, which is how the actual data is encoded on the H1 network.

Figure 1-3 – Manchester encoding.

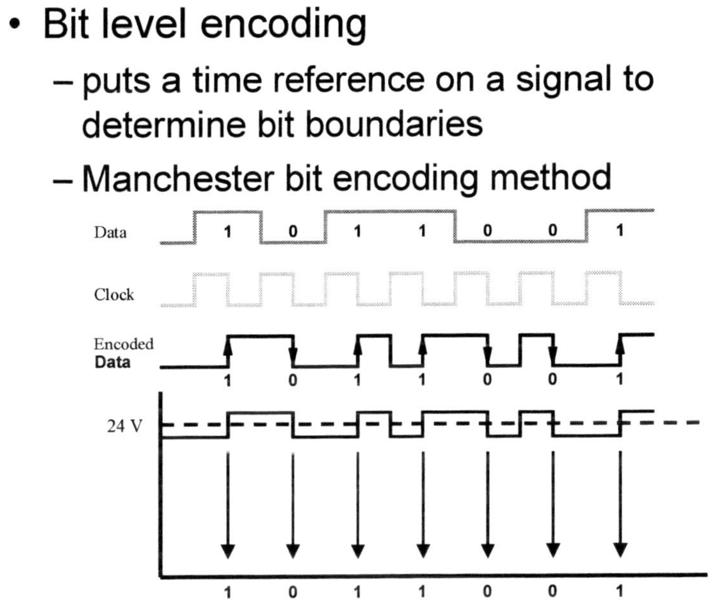

The Data Link Layer (DLL) is a mechanism to transfer data from a node to the other nodes that need the data. The Data Link Layer also manages the priority and order of such transfer requests, as well as data, address, priority, medium control, and other parameters all related to message transfer.

Only one device on a link is allowed to use the medium (Physical Layer) at a time. The Link Active Scheduler (LAS) controls the medium access.

1.1 Topology

The Application Layer provides an interface for the device's application software. This layer defines how to read, write, or start a task in a remote station. The main task of this layer is to define syntax for the messages.

1.1.1 User Layer

The User Layer defines the way of accessing information within Fieldbus devices so that such information may be distributed to other devices or nodes in the Fieldbus network. This is a fundamental attribute for process-control applications.

The architecture of a Fieldbus device is based on Function Blocks, which are responsible for performing the tasks required for the current applications, such as data acquisition, feedback and cascade loop control, calculations, and actuation. Every Function Block contains an algorithm, a database (inputs and outputs), and a user-defined name (the Function Block tag number must be unique in the user's plant). Function Block parameters are addressed in the Fieldbus by means of TAG.PARAMETER-NAME.

A Fieldbus device shall include a defined quantity of Function Blocks.

Function Block. The FOUNDATION™ Fieldbus Function Block, especially its models and parameters–through which you can configure, maintain, and customize your applications–is a key concept of Fieldbus technology.

What is a Function Block? A Function Block is a generalized concept of the functionality in field instruments and control systems such as analog input and output as well as PID control. The FOUNDATION™

Fieldbus specification, "Function Block Application Process–Part 1," gives fundamental concepts, while Part 2 and later give various Function Blocks details.

The Function Block Virtual Field Device (VFD) contains three classes of blocks: Resource Block, Function Block, and Transducer Block.

Resource Block. A Resource Block shows what is in the VFD by providing the manufacturer's name, device name, Device Description (DD), and so on.

The Resource Block controls the overall device hardware and Function Blocks within the VFD including hardware status.

Tip 1 – The mode of the Resource Block controls the mode of all other blocks in the device.

Function Block. A Function Block is a generalized model of measurement and control. The three Function Block classes are:

1. Standard Block, as specified by the Fieldbus Foundation
2. Enhanced Block, with additional parameters and algorithm
3. Open Block or a Vendor-specific Block, designed by individual vendors

Table 1-1 shows these various blocks as defined by the Fieldbus Foundation in the indicated "part" of the specification.

Table 1-1 – Fieldbus function blocks.

Part-2 Blocks	**Standard Blocks**
AI	Analog Input Block
DI	Discrete Input Block
ML	Manual Loader Block
BG	Bias/Gain Station Block
CS	Control Selector Block
PD	P, PD Controller Block
PID	PID, PI, I Controller Block
RA	RATIO Station Block
AO	Analog Output Block
DO	Discrete Output Block
Part-3 Blocks	**Enhanced Blocks**
DC	Device Control Block
OS	Output Splitter Block
SC	Signal Characterizer Block
LL	Lead Lag Block
DT	Dead Time Block
IT	Integrator (Totalizer) Block
(More blocks are under development)	
Part-4 Blocks	**Multiple I/O Blocks**
MDI	Multiple Discrete Input Block
MAI	Multiple Analog Input Block
MDO	Multiple Discrete Output Block
MAO	Multiple Analog Output Block
Part-5 Blocks	
Flexible Function Blocks	IEC61131 Blocks

As Figures 1-4a through 1-4c show, a Function Block has input, output, and contained parameters. Figure 1-4a is a typical Analog Input (AI) Block, 1-4b an Analog Output (AO) Block, and 1-4c a PID Block. Data generated in a block is made available from an output parameter, which can be linked to the input parameter of other Function Blocks.

Figure 1-4a – Analog input block (AI).

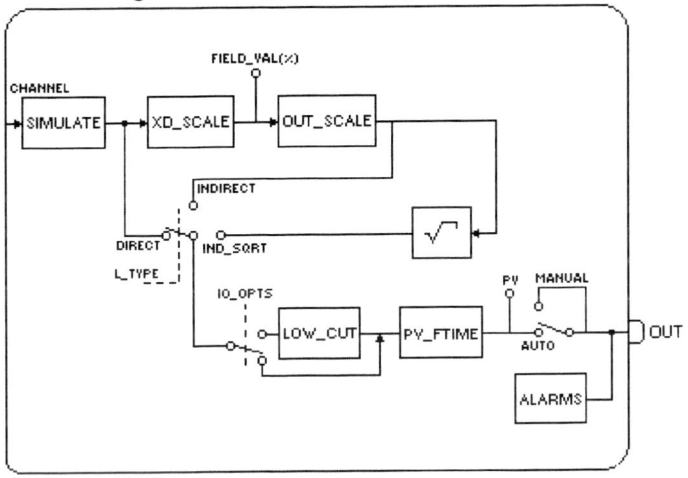

Figure 1-4b – Analog output block (AO).

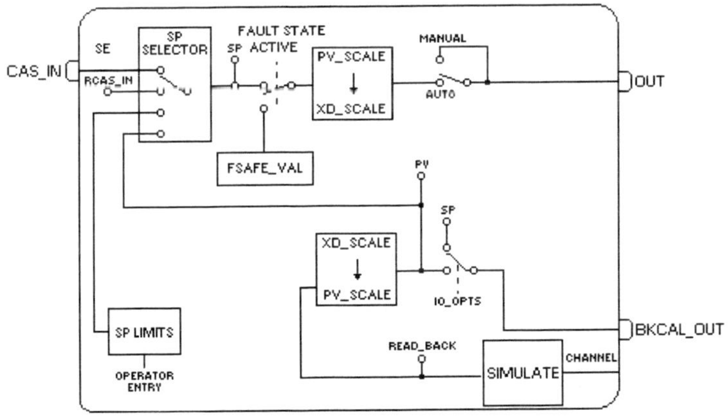

Figure 1-4c – PID block (PID).

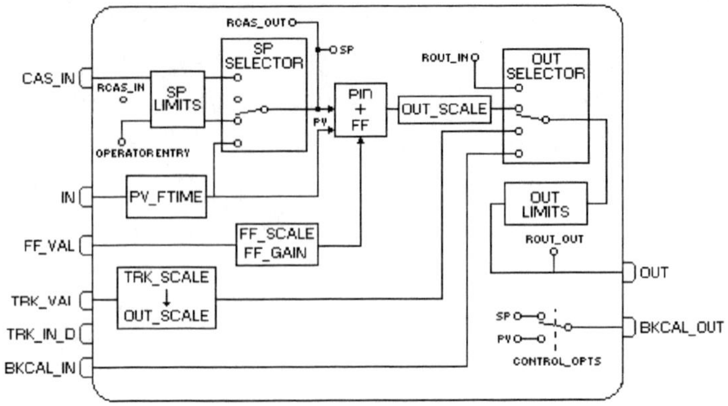

Floating-point parameters have a valid range of ±1.2 × 10^{-38} to ±3.4 × 10^{38}, as well as in some cases the values –Inf or +Inf.

Discrete Blocks have 256 valid enumerated states, which means that in addition to simple logic 0 or 1, they can also be used to represent specific states such as open, closed, true, false, start, stop, running, etc.

In the case of an AO Block, if the actual device reaches either of its open or closed limits, the block will set the corresponding limit in the status element of the associated back-calculation output parameter.

This tells the PID Block to not push the output further in that direction, thus preventing reset windup in the loop.

The operator normally sets the PID loop mode from the Block mode parameter in the PID Block. Control is stopped by setting this parameter to MAN. However, if the operator wishes to "hand operate" the AO Block, it is better to remain in Auto mode and enter the setpoint instead.

Transducer Block. A Function Block is a general idea while the Transducer Block is dependent on its hardware and principles of measurement. For example, a pressure transmitter and magnetic flow meter have different measurement principles but provide an analog measured value. The common part is modeled as an AI Block. The difference is modeled as Transducer Blocks, which provide the information on the measurement principle. A Transducer Block is linked to a Function Block through the CHANNEL parameter of the Function Block.

1.1.2 Physical Layer

This is the Fieldbus layer connected with instrument devices in the field, the standardized speed is 31.25 Kbps; all other speeds shall be used for high-speed interconnection of bridges and gateways (Figure 1-5).

Figure 1-5 – Fieldbus bridge capability.

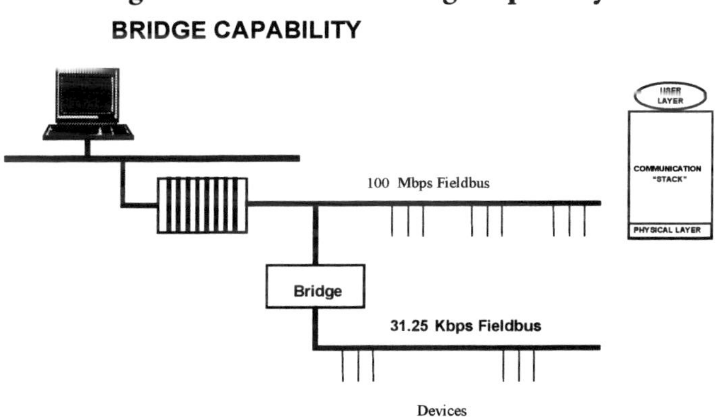

The standard determines the following rules (among others) for the speed of 31.25 Kbps:

1. A Fieldbus instrument shall be able to communicate with the following number of instruments:

   - From 2 to 32 instruments in a non-intrinsically safe connection and power supply separated from the communication wiring

   - From 2 to 6 instruments receiving power by the same communication wiring in an intrinsically safe connection

   - From 1 to 12 instruments receiving power by the same communication wiring in a non-intrinsically safe connection.

Tip 2 – This rule does not forbid the connection of more instruments than the specified number. Such limits have been established considering a consumption of 9 mA ± 1 mA, with a power supply of 24 VDC, and intrinsic safety barriers with an output of 11–21 VDC, and a maximum current of 80 mA for the instruments located within the hazardous area.

2. The length of a bus with the maximum number of instruments operating at a speed of 31.25 Kbps shall not exceed 1900 m in the section of the trunk plus all spurs (Figure 1-6).

Figure 1-6 – Maximum length of Fieldbus network.

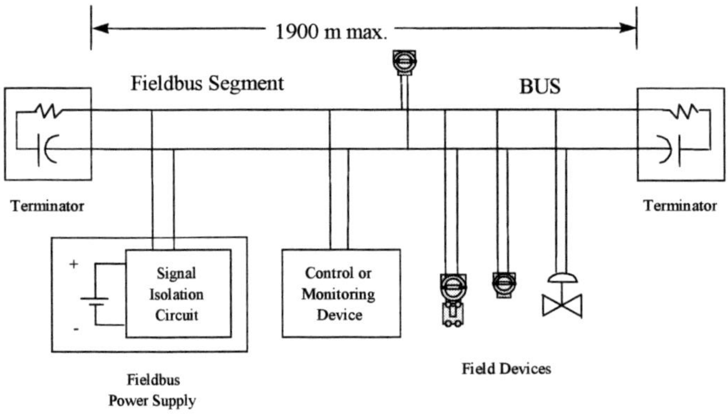

Tip 3 – This rule does not forbid the use of longer lengths, provided that the electrical characteristics of the instruments are observed.

3. The maximum number of repeaters for regenerating the waveform between two instruments cannot exceed four (Figure1-7).

Repeaters are used to expand a Fieldbus network. Repeaters can be either energized or de-energized. When four repeaters are used, the maximum distance between any two devices in a segment is 9500 m.

Figure 1-7 – Fieldbus network with repeaters.

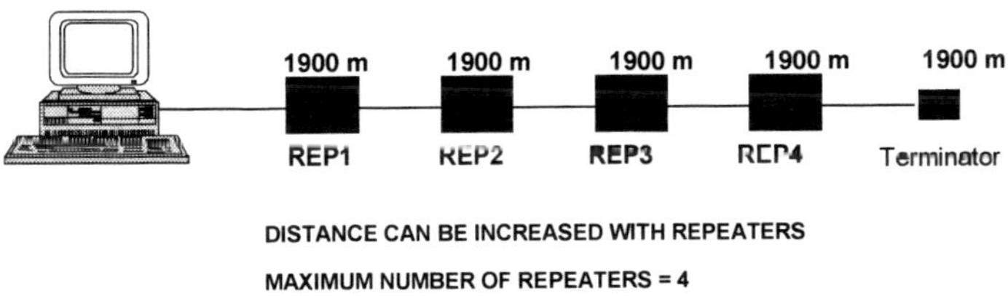

4. A Fieldbus system must be able to remain in operation while a device is being connected to, or disconnected from, it.

5. Failure in any communication element (except for short-circuit or low impedance) shall not affect communication for more than 1 ms.

6. The polarity of systems using twisted pairs shall be respected, their wires shall be identified, and the polarity shall be observed in all connection points. According to the Fieldbus standard, devices themselves are not to be polarity-sensitive, but this is not always the case.

7. For systems with a redundant physical medium:

   - Each channel must comply with the network configuration rules.
   - A non-redundant segment cannot be between two redundant segments.
   - Repeaters must also be redundant.
   - The identification number of the channels must be maintained in the Fieldbus, that is, the Fieldbus channels must have the same numbers as the physical channels.

8. Cable shields shall not be used to conduct power.

Topologies

Several topologies may be used in Fieldbus design. Figure 1-8 shows the topologies that will be detailed below. Power supplies and terminators are not shown so that the figures can be more easily understood.

Bus with Spurs Topology. *Bus with Spurs Topology* uses a single bus to which devices and spurs are directly connected. Several devices may be connected to each spur.

Figure 1-8 – Fieldbus topologies.

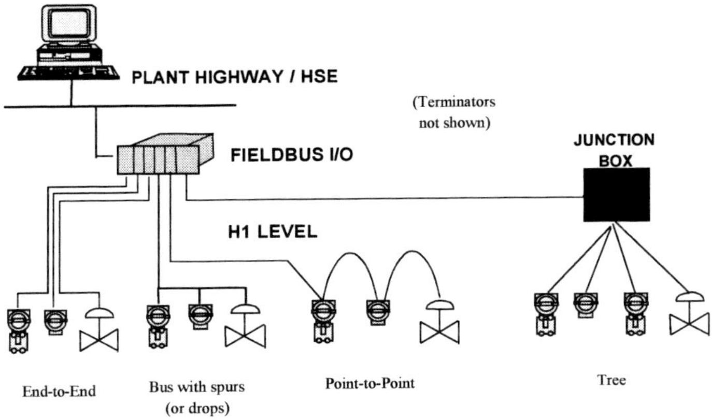

The total spur length is limited according to the number of spurs and the number of devices per spur. This is summarized in Table 1-2 below. The spur table is not absolute. It merely serves as a guideline when designing networks.

Table 1-2 – Maximum spur lengths.

Total devices per network	One device per spur, m (ft)	Two devices per spur, m (ft)	Three devices per spur, m (ft)	Four devices per spur, m (ft)	Maximum total length, m (ft)
1–12	120 (394)	90 (295)	60 (197)	30 (98)	439 (1440)
13–14	90 (295)	60 (197)	30 (98)	1 (3)	384 (1260)
15–18	60 (197)	30 (98)	1 (3)	1 (3)	329 (1080)
19–24	30 (98)	1 (3)	1 (3)	1 (3)	220 (720)
25–32	1 (3)	1 (3)	1 (3)	1 (3)	10 (32)

Bridges are used to connect Fieldbus segments operating at different speeds (and/or physical layers such as wires, fiber optics, radio, etc.)

in order to obtain a large network. A bridge is shown in Figure 1-5 connecting the H1 and High-Speed Ethernet (HSE) networks.

A gateway can be used to connect one or more segments to other types of communication protocols such as Ethernet, RS-232, MODBUS, etc.

Point-to-Point Topology. In *Point-to-Point* or *Daisy Chain Topology,* all devices used in the application are connected in series. The Fieldbus trunk is routed from one device to the next, being interconnected to the terminals of each Fieldbus device.

> ***Caution:*** *Point-to-Point Topology is rarely used since failure of one device in the network will result in total network failure.*

Tree Topology. *Tree* or *Chickenfoot Topology* concentrates the connection of several field devices to couplers/junction boxes. Because of its distribution, the Tree Topology is also known as "Chickenfoot."

End-to-End Topology. *End-to-End Topology* is used to directly connect two devices. The connection may be entirely located at the field (a transmitter and a valve with no other devices connected) or to connect a field device (a transmitter) to the Host Device.

Mixed Topology. *Mixed Topology,* as the name implies, mixes the three most commonly used topologies connected to one another. However, the maximum length of the segment, including the spurs to the total length, shall be observed.

1.2 Communications

Figure 1-9 shows how the registers in a Fieldbus device are "assigned."

Figure 1-9 – FOUNDATION™ Fieldbus node addresses.

0x10 — V(FUN)	Address for Link Master Class devices
0xF7 — V(FUN) + V(NUN)	Address for Basic Class devices
0xF8 — 0xFC	Default address for devices with cleared address
0xFD — 0xFF	Address for temporary devices such as a handheld communicator

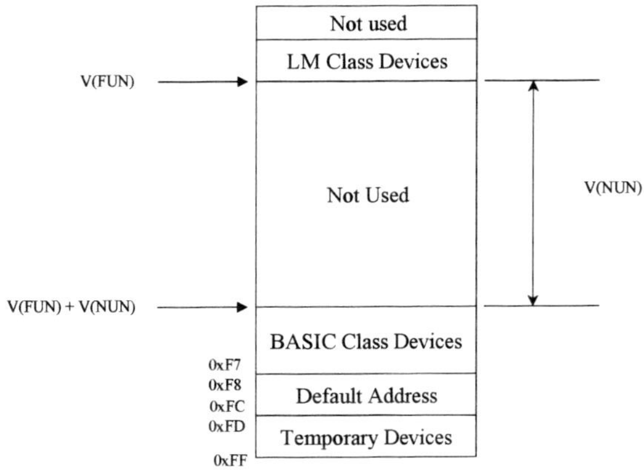

Every Fieldbus device has a unique 32-bit hardware address identifier made up of a 6-byte manufacturer code, a 4-byte device code, and a serial number. This makes it possible to uniquely distinguish each device from the others. The Fieldbus Foundation assigns the manufacturer codes while the manufacturer assigns the device type code and sequential serial number.

A temporary device such as a handheld communicator has a node address in the temporary range. The Link Active Scheduler (LAS) has a node address of 0x04. If a device has an address in the gap V (NUM), it will never be able to join the network. The V (FUN) and V (NUN) parameters are accessible through Network Management.

Several Data Link (DL) addresses are reserved for specific purposes. For example, devices can share the same "global" Data Link Service Access Point (DLSAP) for alarm reception.

Foundation devices are classified into device classes: BASIC, Link Master (LM), and Bridge. An LM class device has the ability to be the LAS, while BASIC class devices do not have this functionality. A Bridge class device has, in addition to LM capability, the functionality to connect networks.

One and only one device in a network can be the LAS at any one time. Therefore at least one LM (or Bridge) class device is needed in a link. LM devices try to acquire the LAS role when no LAS exists on start-up or when the current LAS fails.

Tip 4 –The LM device with the lowest node address becomes the new LAS for the network.

Other secondary or Backup LM devices observe the LAS activity and can assume the primary or Master LAS role if the operating LAS fails.

The LAS manages the scheduled communication part of the synchronized data transfer between Function Blocks.

A Function Block output parameter is a *Publisher* of data and other Function Blocks that receive this data are called *Subscribers*. The LAS controls periodic data transfer from a Publisher to Subscribers using the Network Schedule.

Other network communications take place in an asynchronous way. The LAS is responsible for giving all nodes on a link a chance to send messages.

The third role of LAS is to maintain network communications. The LAS does this by giving the token to all devices detected by the LAS. When a new device is added to the network, it must be recognized by the LAS and added to the token rotation list, which is called the *Live List.*

A Fieldbus device may have user applications that are independent from each other and do not interact. A Fieldbus device consists of Virtual Field Devices (VFD) for such individual applications.

A Foundation device has at least two VFDs. One is the Management VFD where network and system management applications reside. It is used to configure network parameters including VCRs, as well as to manage devices in a Fieldbus system.

The other is a Function Block VFD, where Function Blocks exist. Most field devices have more than two Function Block VFDs.

A measurement or control application consists of Function Blocks connected to each other. Function Blocks are connected through Link Objects in the Function Block VFD. A *Link Object* connects two Function Blocks within a device, or a Function Block to a VCR for Publisher or Subscriber.

A Function Block must get input parameters before its algorithm can be executed. Its output parameters must be published after the algorithm execution. Therefore algorithm execution and Publisher–Subscriber communication must be orchestrated when blocks are distributed among devices. The System Management and Data Link Layer cooperate to achieve this by using the Link Scheduling (LS) time that is distributed and synchronized by the Link Active Scheduler (LAS).

A block has a series of parameters, which have continuous indexes.

1.3 Parameter Classes

Block parameters are classified into three classes: input, output, and contained parameters. Function Blocks can have all of these classes while the Resource Block and Transducer Blocks have only contained parameters.

An *input parameter* is an input of a Function Block and can accept one output parameter of another Function Block. Its data type must be the same as the output parameter.

An *output parameter* is a record consisting of a value (analog or discrete) and its status (showing whether the value is useful).

A *contained parameter* is neither input nor output. It is accessible only through an on-demand *Read* or *Write* request. Its data type can be any of those defined by the Fieldbus Foundation.

Input parameters, output parameters, and some contained parameters are records with an associated status. Status shows whether the value of this parameter is useful or not. If the value is useful, the status is GOOD. If the value is not useful, the status is BAD. The status can be UNCERTAIN when the block is not 100% confident that the value is useful. Blocks have an option to interpret UNCERTAIN as GOOD or BAD.

The acyclical traffic time may be user-defined and configured by means of the configuration software (typical macrocycle times are ≥100 ms).

The above discussion concentrates on communications at the device or H1 level. However, all this data must be shared with other parts of the corporation, including the Distributed Control System[1] (DCS), panel operators via the Human–Machine Interface (HMI), and other parts of the corporation as appropriate. However, the security and integrity of the Supervisory Control and Data Acquisition (SCADA) Layer must be ensured and a variety of methods are available to do so. The most common method, however, is to use firewalls and switches.

This is not different from what is done to protect a corporate local area network (LAN) connected to the Internet through the use of firewalls, switches, and related technologies. The net effect is different "levels" within the enterprise. Figure 1-10 represents this concept.

Figure 1-10 – Layered corporate architecture.

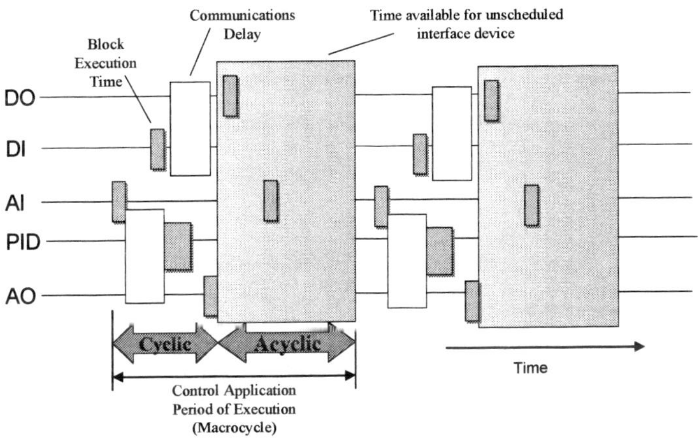

As shown earlier, "layers" and data "paths" are within each of these layers, as well. Data paths are important to prevent information overload to various components or interfaces/devices on the network. The best example of this is device diagnostic information. The panel

1. Historically, DCS has meant Distributed Control System. However in the new era of Digital communications made possible with fieldbus technology the more appropriate term is Digital Control System and hence this term will be used for this text.

operator needs to know only the process measurement and its status; hence all the other information should be routed through a separate system that can be better integrated with a facility's computerized maintenance system. Figure 1-11 shows how these two data paths may exist on the same Control/SCADA Layer. This data "splitting" is possible because Fieldbus uses Publisher–Subscriber relationships to transfer data between memory registers/devices. In this case, the process data is subscribed to by the operator interface while the diagnostic data is subscribed to by the Condition Monitoring System.

Figure 1-11 – SCADA layer Fieldbus traffic "management."

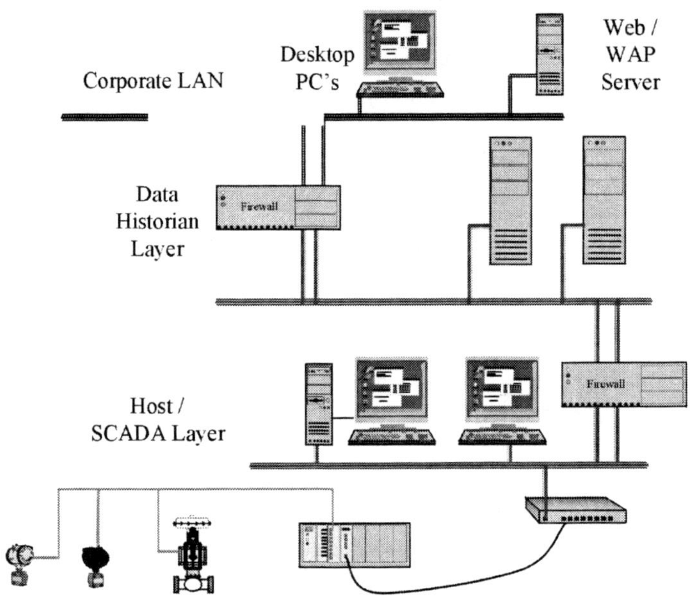

How should all this new information be used? There is a great deal of activity under way on a number of fronts related to, but not part of, the Fieldbus effort that will enable the integration of this information into the larger corporate network environment. Some of these initiatives include:

- OPC–OPen Connectivity technology, as developed by the OPC Foundation (www.opcfoundation.org) and formerly known as Object Linking and Embedding (OLE) for Process Control, is a technology to enable real-time transfer of data from one system to another through the use of standardized call functions. The OPC Foundation is developing this open standard.
- XML–eXtensible Mark-up Language is the most complex form of "language" used to display information via the Web. XML is object-based and therefore each piece of information can potentially be self-defined, thus making it viewable by anyone anywhere.

2 – Fieldbus Cabling

With an understanding of what all the "pieces" are, it is time to start "putting them together." Chapter 2 summarizes information on FOUNDATION™ Fieldbus physical layers, H1 and HSE, including information on sizing and connecting all the pieces together to build a network.

The FOUNDATION™ Fieldbus H1 Network uses twisted-pair wires. A twisted pair is used, rather than a pair of parallel wires, to prevent external noise from getting onto the wires. A shield over the twisted pairs further reduces the noise.

For new installations or to get maximum performance for a FOUNDATION™ Fieldbus Network, twisted-pair cable designed especially for FOUNDATION™ Fieldbus should be used. The important twisted-pair cable characteristics are listed in Table 2-1.

Table 2-1 – Fieldbus cable characteristics.

Type	A	B
Wire size	18 GA (0.8 mm)	22 AWG (0.32 mm)
Shield coverage	90%	90%
Attenuation at 39 kHz	3 dB/km	5 dB/km
Characteristic impedance at 31.25 kHz	100 Ω ± 20%	100 Ω ± 30%
Capacitance	2 nF/km	2 nF/km
Resistance	44 Ω/km	112 Ω/km
Maximum propagation delay between 0.25 f_r and 1.25 f_r	1.7 µs/km	1.7 µs/km

In new installations, a type A twisted-pair shielded cable shall be specified. Other types of cables may be used, provided the limits presented in Table 2-2 are observed, such as twisted-pair multicables with an overall shielding called type B cable.

Table 2-2 – Cable type specifications.

Parameters	Conditions	Type A	Type B	Type C	Type D
Characteristic impedance, Z_0, Ω	f_r (31.25 kHz)	100 ± 20	100 ± 30	**	**
Maximum DC resistance, Ω/km	per conductor	22	56	132	20
Maximum attenuation, dB/km	1.25 f_r (39 kHz)	3.0	5.0	8.0	8.0
Nominal cross-sectional area of the conductor (gauge), mm^2		0.8 (#18 AWG)	0.32 (#22 AWG)	0.13 (#26 AWG)	1.25 (#16 AWG)
Max. non-balanced capacitance, pF	1 m long	2	2	**	**
Maximum length, m		1900	1200	400	200
Capacitance, pF/m		150	150	150	
Attenuation, dB/km		3	5	5	

** not specified.

A type of cable that is not recommended is the single or multiple twisted pair without shielding called type C cable.

The least used type is the multipair cable without twisted pairs and with overall shielding called type D cable.

Types C and D cables are listed in the specification for completeness and so users are aware of the cables' limitations if they choose to use them as part of a retrofit project.

The maximum length indicated on the specifications are recommendations that include a safety factor to satisfactorily reduce communication problems. As a general rule, the maximum cable run is basically related to the cable type and its characteristics, the chosen topology, and the quantity and type of devices used (Table 2-3).

Table 2-3 – Cable types and their maximum lengths.

Type	Description	AWG	Capacitance pF/m	Attenuation, dB/km	Max length, m
X_o	Multiconductor with overall shield	20	75	4	1200
X_i	Multiconductor with individual and overall shield	20	98	5	1900
X_s	Single pair	11	44	6	1900

The maximum overall length of cable when mixing cable types is determined according to the formula

$$\frac{L_x}{L_{max_x}} + \frac{L_y}{L_{max_y}} \leq 1$$

where:

L_x = length of cable x

L_y = length of cable y

L_{max_x} = maximum length of cable type x alone

L_{max_y} = maximum length of cable type y alone

A FOUNDATION™ Fieldbus Network consists of the total wire pair length, trunk length, and spur lengths. Total wire pair length is the sum of the trunk cable plus all the spurs. A trunk is the longest cable run between any two devices on a network. A spur can vary in length from 1 to 120 m. A spur that is < 1 m is considered a splice.

The same ratio calculation can be performed if more than two cable types are present in the system, as long as the sum of the ratios is < 1.

In addition to the physical limitations described earlier, which are generic guidelines based on voltage and capacitive limitations, the following equation can be used to calculate the maximum trunk length on a system with approximately equal spur lengths and devices with nearly equivalent current, voltage, and capacitance needs.

$$L_{TMax} < \frac{((V_{PS} - V_{Min}) \times 10^6 - I_D \times 2 \times R_S \times L_S)}{\sum I_D \times 2 \times R_T}$$

where:

L_{TMax} = maximum voltage of trunk cable, meters
V_{PS} = power supply voltage, volts
V_{Min} = largest minimum voltage of all the field devices, volts
I_D = DC current draw of the field device with the largest minimum voltage, milliamps
R_S = manufacturer-specified resistance of spur cable, Ω/km
L_S = length of spur cable, meters
ΣI_D = total (sum) of DC current draw of ALL field devices, milliamps
R_T = manufacturer-specified resistance of trunk cable, Ω/km

If the installation is a Chickenfoot arrangement, or if each of the field devices has very different minimum voltages (e.g., a temperature transmitter and a valve positioner) and current specifications, then the voltage available at each device on the segment should be calculated using the following formula:

$$V_D = V_{PS} - (\sum I_D \times 2 \times R_T \times L_T + I_D \times 2 \times R_S \times L_S) \times 10^{-6} > V_{Min}$$

where:

V_{Min} = minimum voltage of the field devices, volts
V_D = DC voltage available at the field device, volts
V_{PS} = power supply voltage, volts
ΣI_D = total (sum) of DC current draw of ALL field devices, milliamps
R_T = manufacturer-specified resistance of trunk cable, Ω/km
L_T = Length of trunk cable, meters
I_D = DC current draw of the field device, milliamps
R_S = manufacturer-specified resistance of spur cable, Ω/km
L_S = length of spur cable, meters

Capacitance constraints must also be considered since the effect on the signal of a spur $<$ 300 m long is very similar to that of a capacitor. In the absence of actual data from the manufacturer, a value of 0.15 nF/m can be used for Fieldbus cables.

$$C_T = \sum (L_S \times C_S) + C_D$$

where:

C_T = total capacitance of network, nF
L_S = length of spur cable, meters

C_S = Capacitance of wire for segment, nF/m (use 0.15 if no other number is available)

C_D = Capacitance of device, nF

The attenuation associated with this capacitance is 0.035 dB/nF. To estimate the attenuation associated with the installation, the following formula provides a useful guideline.

$$A = C_T \times L_T \times 0.035 \frac{dB}{nF} < 14dB$$

where A is Attenuation, dB.

Tip 5 – Total attenuation must be < 14 dB, which is the minimum signal that can be detected between the lowest level transmitter and the least sensitive receiver.

Connector Blocks are optional devices used in installations where the devices must be periodically disconnected and/or displaced. They can also be very convenient for connecting temporary equipment to a certain point. Figure 2-1 shows a number of these connections.

The same Terminal Block used in 4- to 20-mA applications can be used for Fieldbus, if connected as shown in Figure 2-2.

Terminal Blocks make multiple trunk connections possible, so that equipment can be connected to any terminal set in the trunk.

Figure 2-2 shows how a Fieldbus segment termination and connection to several field devices in a junction box could be done with conventional Terminal Blocks.

Figure 2-1 – Fieldbus connector blocks.

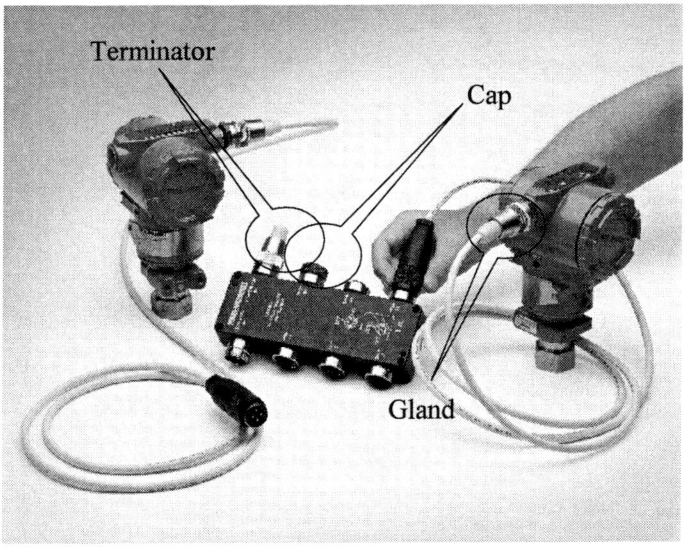

Figure 2-2 – Terminator inside a junction box.

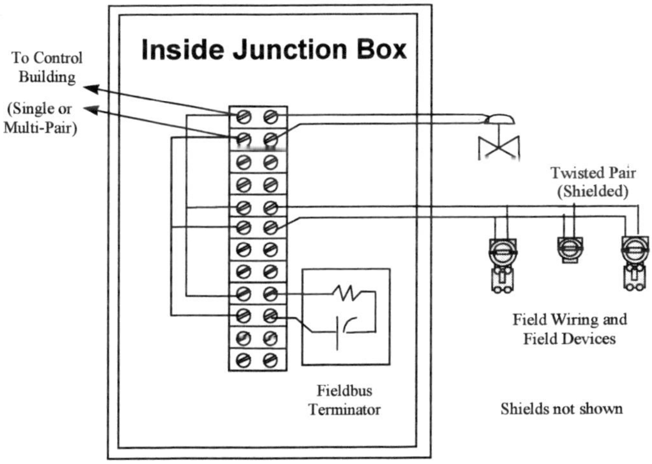

Tip 6 – The terminator should be placed in the junction box so that accidental removal of a field device does not result in loss of network communications.

Tip 7 – If a multiple home-run cable goes to a field junction box, do not attach the cable shield wires from different networks together. This creates ground loops and introduces noise into the network.

Do not connect the cable shield of a field instrument to the instrument ground or chassis.

Figure 2-3 illustrates the correct way to terminate a field device. Note that special care must be taken to ensure isolation of the shield and signal cables since this can introduce extraneous, sporadic noise to the network.

Figure 2-3 – Termination guidelines.

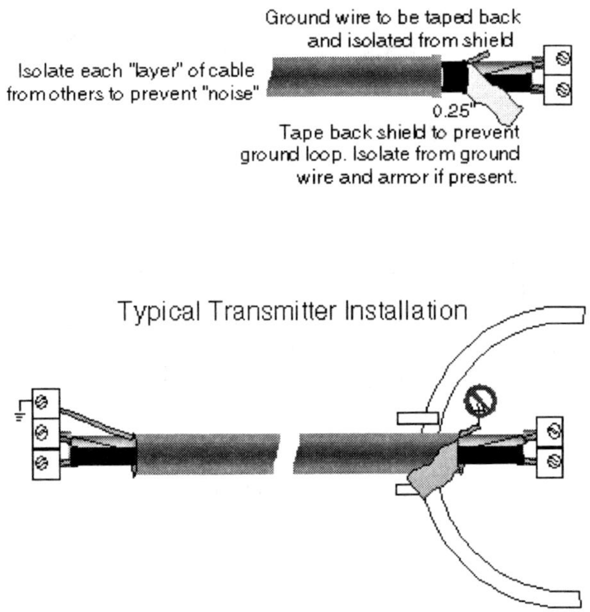

The illustrated termination can be achieved by adhering to the following good installation practices:

- Use proper wire strippers that do not nick the wire as they strip the insulation.
- Use color-coded crimp ferrules to prevent stranded wires from getting loose and shorting to other wires. Make sure ferrule material is the same as the wire material on the wiring blocks. Similar metals are more corrosion-resistant than a bare wire in a wire terminal.

The following represents a series of guidelines to consider when designing FOUNDATION™ Fieldbus segments. In some cases, the guidelines may conflict with each other or an economic design decision made by the project team.

- Each segment will normally have eight or fewer devices and not more than two control valves.
- A general rule to remember is that "shorter is better" when it comes to designing spurs.
- To allow for devices to be added to the network in the future if required, networks shall be designed for 25% spare capacity. Remaining capacity shall be calculated on the basis of the lesser of the power or bandwidth constraint.
- A Fieldbus segment in a hazardous area shall consist only of the type and number of devices that will not cause the segment current draw to exceed the rating of the intrinsically safe (IS) barrier or IS power supply (typically 80 mA).
- For pressure vessels with multiple temperature and/or pressure measurements, the multiple measurements should be distributed to reside on separate H1 cards.

2.1 High-Speed Ethernet

High Speed Ethernet has been designed to fully incorporate all the functionality of Fieldbus H1 and supports both the Transmission Control Protocol (TCP) and User Datagram Protocol (UDP) Transport Layer protocols, but UDP is the default transport protocol. In addition, HSE supports full redundancy of devices and the Data Link Layer.

Figure 2-4 shows how the fieldbus-specific and Commercial-Off-the-Shelf (COTS) parts of the specification combine to provide this integrated solution. In this figure, the shaded areas are defined by other than the Fieldbus Foundation and are common to many of the other Industrial Ethernet protocols. The balance of the "blocks" in the diagram is defined in the Fieldbus protocols prepared as part of the HSE specifications and should only be of interest to a device manufacturer. The key blocks to allow encapsulation of the H1 protocol within HSE are the H1 Interface blocks 1 through N for each H1 function block to be connected to the HSE network.

Figure 2-4 – HSE profile functional areas.

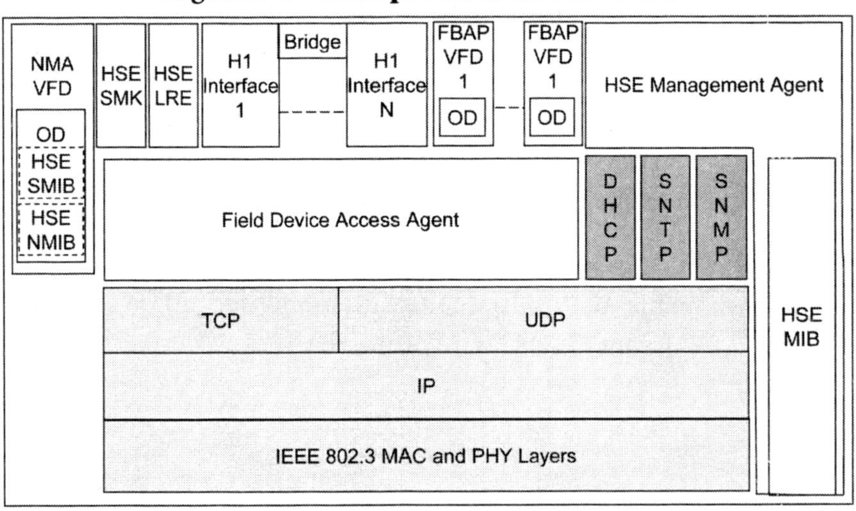

All Ethernet devices are built on the same platform. Therefore they can coexist on the same network with different Application Layer protocols without conflict. Since both Ethernet and Internet Protocol (IP) standards have addressing schemes called Ethernet Medium Access Control (MAC) address, which is a hardware address and an IP address, it is possible to mix devices from Profibus and Fieldbus on the same physical wires. The MAC address is set in the hardware by the device manufacturer, and it is unique to each type of device.

Although possible, however, it is not recommended to share the physical layer among different protocols and place the Industrial Ethernet protocol, Corporate LAN traffic, and even VoIP phones on a single wire or fiber. If you decided to try this, the network as a minimum must use IP V6 switches to not only minimize the impact of any data storms but also monitor each packet for its priority so that the control-related information always had the highest priority. However, since the most significant cost of installing cable is not the cable itself but the associated labor, there is no excuse for doing so.

When preparing the HSE specifications, the Foundation made an effort to determine the types of functionality that would be required of devices connected to it and came up with the following set of requirements or profile classes:

- Four types of Field Devices identified as Class 41
- Four types of Linking Devices identified as Class 42
- An I/O Gateway Device, Class 43
- Two classes of Host Devices, Classes 44 and 45
- Two levels of redundancy, Classes 46 and 47 and Classes 48 and 49, which are used for timing, synchronization, and network connectivity.

The following table summarizes the differences between the various classes of device.

Table 2-4 – HSE class summary.

Class	Main Feature
41a	Function Blocks, sources events, publishes data, responds to client requests
41b	41a plus Subscribes
41c	41b plus basic FFB support
41d	41c plus extended FFB support
42a	Linking Device - Client access HSE ⇨ H1, server response H1 ⇨ HSE (e.g., configuration)
42b	42a plus sources events from H1 ⇨ HSE
42c	42b plus publisher/subscriber H1 ⇔ H1, H1 ⇔ HSE
42d	42c plus extended services for Flexible Function Blocks in H1 devices
43	I/O Gateway device - allows interoperation with other instrument systems
44	Simple Host
45	Configurator Host Reserved
46-D1	Does not support redundancy, but sends required redundancy diagnostic messages
46-D2	46-D1 plus redundant device - manual switchover
46-D3	46-D2 plus automatic switchover
47-L1	Single HSE communication port
47-L2	Dual HSE communication ports
48-TC	HSE Time Synchronization - can request time from server
48-TS	48-TC plus can be the time server
48-TN	Has no time synchronization capabilities
49-AY	Automatic IP Address - using DHCP
49-AN	IP Address is obtained using local means

A Class 43 I/O Gateway Device supports Fieldbus Messaging Services (FMS) that are supported by Class 42c.

A Class 44 Simple Host is an HSE host of some kind; for example, it might be a Process Operator Workstation that supports subscription to Function Block data such as AI, AO, PID etc. and can be a report sink or receiver of data from the network.

The Class 45 Configurator Host is as the name implies–an HSE host with configuration capabilities. It is capable of configuring HSE and H1 devices and configuring HSE LAN Redundancy Information in HSE Devices.

Every HSE device is also required to support as a minimum Classes 48 and 49 and one of Classes 46 and 47.

For 10 Mbps Ethernet, there can be no more than four repeaters or shared hubs between any two devices on the network. The hubs can typically be located at a maximum distance of 100 m from each other. Foundation Fieldbus uses 100 Mbps full-duplex standard Ethernet as its basis so some of the limits of 10 Mbps Ethernet are overcome though not the ones listed above.

Ethernet can be made "deterministic," meaning the message will be delivered within a specified time frame by using 100Base-TX and switched hubs with a maximum 50% loading, and passing the message through the minimum number of nodes on the network. This way, the chance of a message getting delayed is less than the chance of a message being lost due to noise and the network limitation resides in the switch not the other system components.

To make networking easier to administer, it is a good idea to use patch panels at the hub to facilitate interconnections between various networks and sub-networks. Patch cords are then used to make the desired final connections.

Two types of twisted-pair cables are used for Ethernet:

1. UTP, which has four *unshielded twisted pairs*, sometimes with an overall shield. This is the preferred cable for industrial Ethernet installations.
2. STP, which has four individually, *shielded twisted pairs* with an overall shield called Category 7 cable (rarely used.).

Only two of these pairs are used for 10Base-T and 100Base-TX. Twisted-pair cables use RJ45-style connectors, and industry is working to develop a ruggedized form of this termination assembly.

UTP cable should never be placed in a metal conduit, to prevent degraded performance. If it is necessary to run the cable in a harsh environment, use a special cable with higher pull strength and suitable jacket for the conditions.

Figure 2-5 shows how a twisted pair is used for Ethernet connectivity, while Figure 2-6 shows how HSE is connected to the H1 network to create a full control network.

Since a process-control network is typically not connected directly or permanently to the Internet, but rather through an internal gateway between levels, one of the three free IP network numbers is generally used for the host-level network, and an IP address from the associated blocks of address space is assigned to each device on what is, in effect, a private network. The gateway performs Network Address Translation (NAT) effectively isolating all automation devices behind the Gateway, which is a proxy server for these devices. The Gateway can also support strong user authentication to gain access to automation devices.

Figure 2-5 – Ethernet wiring.

Physical Media

Name	Cable	Max Segment	Nodes/ Segment
10Base5	Thick Coax	500 m	100
10Base2	Thin Coax	200 m	30
10Base-T	Twisted pair	100 m	1024
10Base-F	Fiber	2000 m	1024

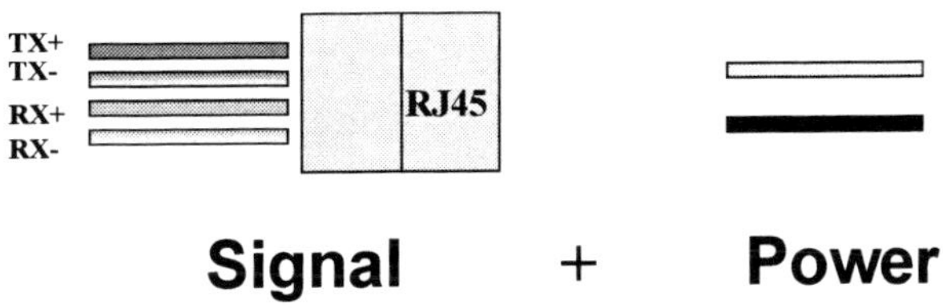

Figure 2-6 – Wiring H1 devices to a linking device.

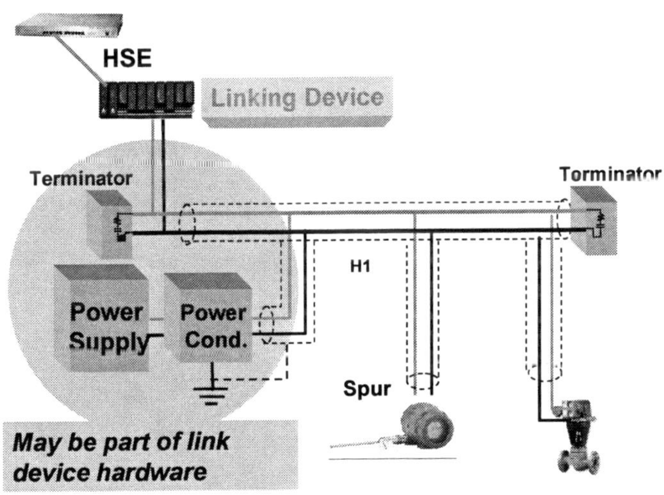

Tip 8 – The decision regarding the allocation of IP addresses must be shared and, if necessary, negotiated with the corporate information technology department to ensure

there is no potential for conflict once communications between the two networks are established.

Because Ethernet standards do not directly support ring topology, if it is decided that a Ring Topology is to be used for the Ethernet backbone, all the hubs/switches/repeaters that are part of the ring must be from the same manufacturer.

Fieldbus HSE is another link in the integrated digital enterprise of the future as shown in Figure 2-7. H1 and in some cases HSE Field Devices will be used in the field, HSE will be used with Linking Devices to gather multiple field level signals onto the Control network and then conventional Ethernet with appropriate security.

Figure 2-7 – Integrated fieldbus system.

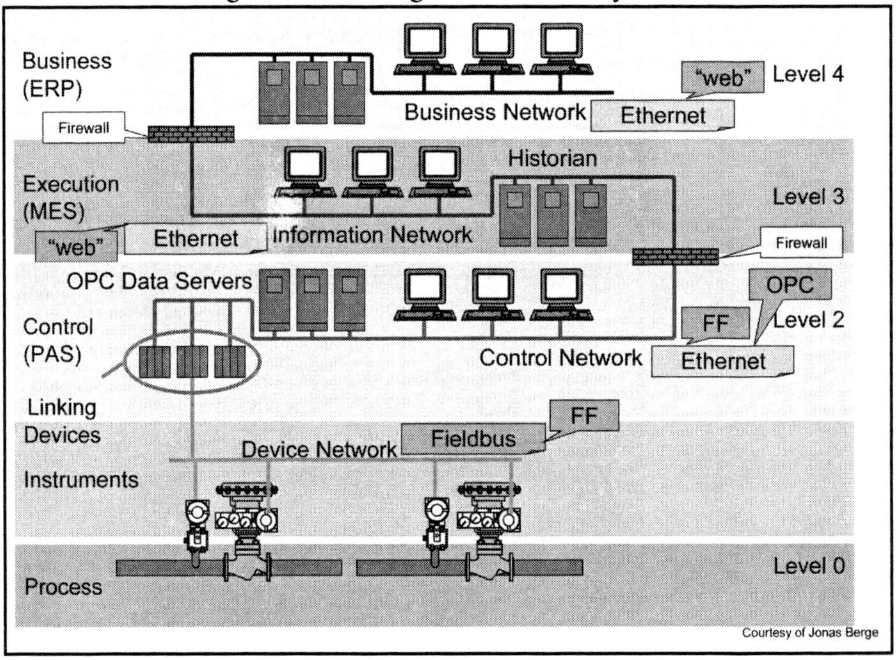

3 – Fieldbus Power Supplies

As seen in Chapter 2, FOUNDATION™ Fieldbus was designed to use as much of the existing infrastructure as possible. This means that it can also be installed in the same environments. However, because Fieldbus is a digital communication network, it has unique power conditioning requirements. This chapter discusses those power requirements, for three different conditions; Intrinsically Safe (IS), Non-Intrinsically Safe (NIS) and the new Fieldbus Intrinsically Safe Concept (FISCO) systems.

Fieldbus devices may receive their power supply either by the same communication cables or separately.

Table 3-1 defines the characteristics required of network or loop powered devices. While Table 3-2 lists the requirements for the associated power supply for these devices.

Table 3-1 – Characteristics of network energized devices.

Characteristics of network-powered devices	Limits for 31.25 Kbps
Operating voltage	9.0 to 32.0 VDC
Maximum voltage	35 V
Maximum idle current rate of exchange (not transmitting); this is not applicable to the first 10 ms after the device is connected to an operating network, nor in the first 10 ms after the network is energized.	1.0 mA/ms
Maximum current: This is adjustable in the interval of 100 µs to 10 ms after the device is connected to an operating network, or 100 µs to 10 ms after the network is energized.	Idle current plus 10 mA

Table 3-2 – Networks' power supply requirements.

Networks power supply requirements	Limits for 31.25 Kbps
Output voltage, non-intrinsically safe	≤32 VDC
Output voltage, intrinsically safe (IS)	Depends on the barrier
Output impedance, non-intrinsically safe, measured within the frequency range of 0.25 to 1.25 f_r	≥3 kΩ
Output impedance, intrinsically safe, measured within the frequency range of 0.25 to 1.25 f_r	≥400 kΩ (An intrinsically safe power supply includes an intrinsically safe barrier.)

If an ordinary power supply were to be used to power the Fieldbus, it would absorb the signals on the cable because it would try to maintain a constant voltage level. Fieldbus power supplies are "conditioned" by putting an inductor be-tween the power supply and the field cable, and adding a resistor to the inductor to prevent "ringing." The power conditioner consists of the equivalent of a 50-Ω resistor and 5-millihenry inductor in series.

The Fieldbus Foundation released a specification, FF-831 Fieldbus Power Supply Specification, in 2004. All Fieldbus Foundation Power Supplies/Conditioners used in a project should be tested against this specification and have received the Fieldbus Foundation certification "check mark."

These parallel shunts result in a 50-Ω impedance, and as the transmitter's Manchester current passes through, it creates a 0.5-VAC voltage drop. All the devices on the network are sensitive enough to pick up the attenuation of the signal along the wires. The net effect is that the transmission is done as current and the reception as voltage.

All Fieldbus systems may or may not be sensitive to polarity. Considering the Physical Layer standards, there is no requirement for polarity insensitivity. The use of nonpolarized equipment, however, simplifies the design and the installation.

Tip 9 – Not all manufacturers are polarity-insensitive so the user must be cautious when specifying and purchasing devices.

A Fieldbus cable shield, by usual practice, is grounded at one end only, and it cannot be used as a power conductor.

Fieldbus devices connect to the network in parallel, and to be truly specification-compatible, they should be polarity-insensitive. Despite the fact that Fieldbus devices are to be polarity-insensitive, in many cases this is not true; so care must be taken when installing devices to match the polarity indicated on the device terminals to that of the wire pair from the host and power supply. For this reason, it is a good idea to continue to use the wire pair colors that the on-site maintenance people are used to.

In the case of shielded cables, the impedance measured between the Fieldbus cable shield and the Fieldbus device ground shall be >250 kΩ for all frequencies below 63 Hz.

The maximum nonbalanced capacitance for the ground of both input terminals of a device shall not exceed 250 pF.

A terminator shall be placed on both ends of a trunk cable, connecting one signal conductor to another. No connection can be done between the terminator and the cable shield.

The terminator impedance shall be 100 Ω ± 20% within the frequency range of 0.25 to 1.25 f_r (7.8 to 39 kHz). This is approximately the mean impedance of the cable in the operating frequencies and has been chosen to minimize reflections in the transmission line.

Direct current leakage by the terminator shall not exceed 100 µA. The terminator shall be nonpolarized.

To comply with noise immunity requirements, it is necessary to make sure there is shielding continuity along the cabling, connectors, and couplers.

Fieldbus devices are specified to work with voltages of 9–32 VDC.

Tip 10 – The 9 VDC specified is a minimum; it is highly desirable that a margin of at least 1 V (i.e., a minimum of 10 VDC) be maintained. Some devices do not conform to FF standards and require 11 V to operate. Any segment designed to operate below 15 V normally should carry a warning about additional loads in the segment documentation. Minimum segment voltage should always be shown in the segment documentation.

3.1 Intrinsic Safety

Intrinsic Safety (IS) is an instrumentation design methodology for flammable atmosphere areas. Safety is obtained by limiting the power and current values that could create sparks or cause surface heating as a result of their normal operating conditions or electrical charges that could cause an ignition. Flammable materials may ignite because of two unrelated parameters: the minimum quantity of energy required to create sparks capable of igniting a given flammable gas, and the minimum temperature at which a heated surface will cause the same effect.

The impedance of intrinsic safety barriers shall be higher than 400 Ω for all frequencies in the interval between 7.8 and 39 kHz. An intrinsic safety barrier cannot be more than 100 m away from the terminator. The terminator resistance shall be sufficiently low so that when it is in parallel with the barrier's impedance, the equivalent impedance is an entirely resistive one.

This requirement is valid for separate intrinsic safety barriers and for those internally integrated to the power supplies.

Within the operating voltage range of an intrinsic safety barrier (in the 7.8–39 kHz interval), the capacitance measured between the positive terminal (hazardous side) and the ground shall be less than 250 pF higher than the capacitance measured between the negative terminal (hazardous side) and the ground.

Tip 11 – In the case of intrinsically safe systems, the operating voltage may be limited to comply with the certification requirements. In this case, the power supply will be located within the safe area, and its output voltage will be attenuated by a safety barrier or by an equivalent component.

In Europe, equipment is classified on the basis of its design and construction characteristics; in North America, equipment is classified considering its possible area of installation. In practical terms, both systems are equivalent if the differences are smaller than those presented in Table 3-3.

As a general rule of thumb, if the device characteristics are not known, a good starting point for designs is shown in Table 3-4.

Table 3-3 – Equipment classification guide.

Explosive substance	Equipment classification		Ignition power
	Europe	North America	
Methane	**Group I**	**Non-classified**	
Acetylene Hydrogen Ethylene Propane	Group II, C Group II, C Group II, B Group II, A	Class I, group A Class I, group B Class I, group C Class I, group D	>20 μJoules >20 μJoules >60 μJoules >180 μJoules
Metal dust Coal dust Grain dust	Under elaboration	Class II, group E Class II, group F Class II, group G	Easier Ignition
Fibers		Class III	

Table 3-4 – Default function block information for designs.

Function Block	Execution time (ms)	Current (mA)
AI	30–50	15–20
AO	60–75	25–30
PID	30	N/A

3.2 Fieldbus Intrinsically Safe Concept

The Fieldbus Intrinsically Safe Concept (FISCO) was developed to provide a way to supply additional power to the Fieldbus segment while still keeping the energy level below that which could cause an explosion.

Figure 3-1 – Protective systems incendive limits.

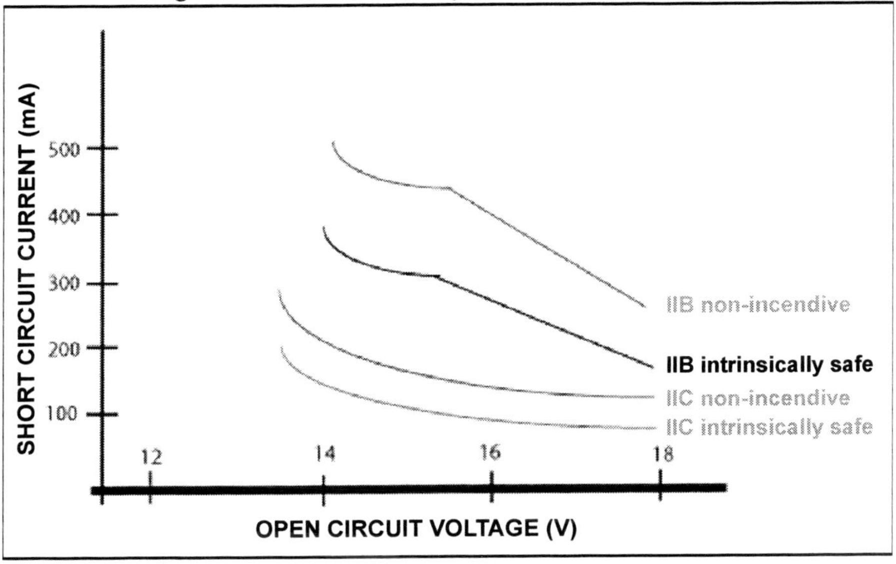

The intrinsic safety of Fieldbus installations based on the Physical Layer standard IEC 61158–2 requires special consideration. Because of the limited DC power to be shared by a number of field devices, long cable runs, and terminators storing capacitive energy, the traditional intrinsic safety installation and interconnection rules restrict the application of such systems. FISCO allows the interconnection of an intrinsically safe apparatus to an associated apparatus not specifically examined in such combinations. The criterion for such interconnection is that the voltage (Vi), the current (Ii), and the power (Pi) that the intrinsically safe apparatus can receive and remain intrinsically safe, considering faults, must be equal to or greater than the voltage (Vo), the current (Io), and the power (Po) that can be delivered by the associated apparatus (supply unit). In addition, the maximum unprotected residual capacitance (Ci) and inductance (Li) of each apparatus (other than the terminators) connected to the Fieldbus must be =5 nF and 10 µH, respectively.

In each IS Fieldbus segment, only one active device, normally the associated apparatus, is allowed to provide the necessary power for the

Fieldbus system. The allowed voltage (Vo) of the associated apparatus used to supply the bus is limited to a range of 14 VDC to 24 VDC.

All other equipment connected to the bus cable must be passive, meaning that the apparatuses are not allowed to provide energy to the system, except a leakage current of 50 μA for each connected device.

Figure 3-2 – Typical FISCO network.

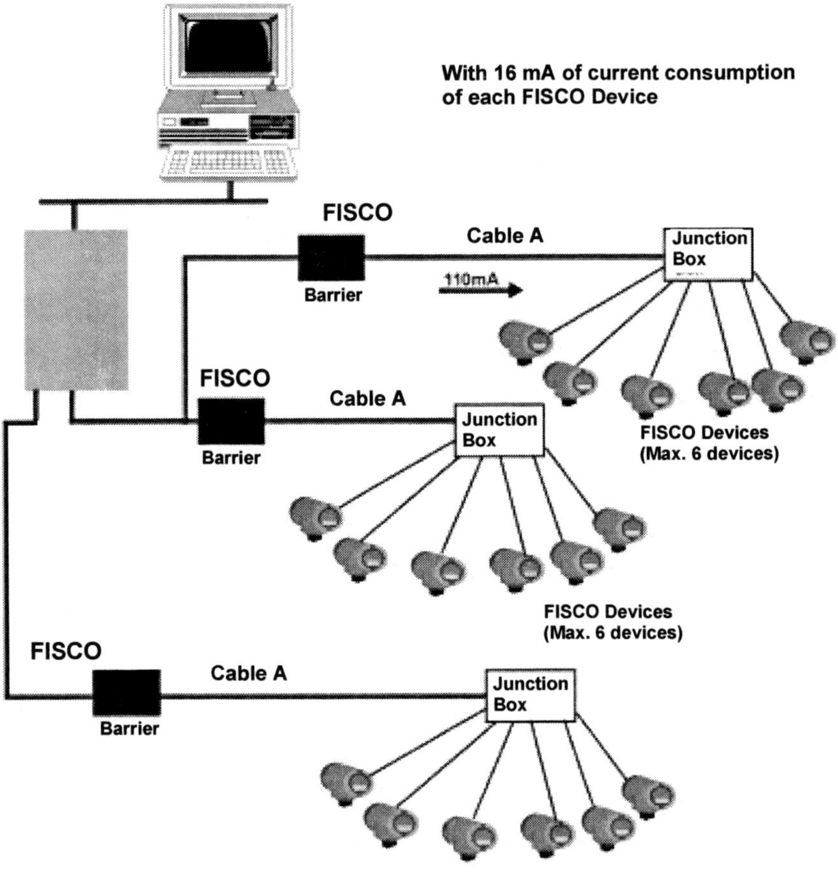

Table 3-5 summarizes the FISCO parameter range as defined in the Physikalisch-Technische Bundesanstalt (PTB) Report PTB W39.

Table 3-5 – FISCO parameters.

	Area Classification	
	Eex ia IIC	**Eex ib IIC / IIB**
Supply unit	Trapezoidal output characteristic	Approximately rectangular output characteristic
V_S (IS maximum value)	14–24 V $V_o \geq 2 \times V_S$	14–24 V
I_k	According to PTB report W-39	According to PTB report W-39
Cable parameters/km		
R'	15–150 Ω	15–150 Ω
L'	0.4–1 mH	0.4–1 mH
C'	80–200 nF (incl. screen, if existing)	80–200 nF (incl. screen, if existing)
$C' = C'_{conductor/conductor} + 0.5\ C'_{conductor/screen}$ if the bus circuit is potential-free (balanced)		
$C' = C'_{conductor/conductor} + C'_{conductor/screen}$ if the screen is connected with one pole of the supply unit.		
Maximum cable length if there are no safety-related restrictions.		1000 m
Terminators	RC elements with	
R	90–100 Ω	
C	90–100 μF	
Note: One terminator at each end of the trunk cable is permissible. The resistor must be infallible according to EN 50020.		

Outro conceito que também está surgindo é o FNICO (Fieldbus Nonincenditive Concept) que é um conceito similar ao Fisco mas restrito a uso em Zona 2 .

Table 3-6 – Characteristics of FISCO and FNICO networks.

Type of Power Supply			FISCO		FNICO	
Apparatus Group			IIC	IIB	IIC	IIB
Safety Description	Voltage	V	14.0	14.8	14.0	14.8
	Current	mA	180	359	233	380
Useable Output	Voltage	V	12.0	12.8	12.0	12.8
	Current	mA	110	250	180	320
Maximum number of devices	Assumes 15mA per device		7	16	12	21
Maximum length of trunk for 7 devices	Assumes 50 ohms per Kilometer and 15mA per device		560m	690m	560m	690m

As with any design, one of the key considerations to good design is risk management. Since Fieldbus places multiple devices on a single, highly reliable segment, there is always the small risk that the segment and all the devices on it could fail. For this reason, all your control loops should be rated on a criticality level.

3.3 Fieldbus Non-Incendive Concept

FNICO is a derivative of FISCO and is specifically intended for fieldbus installations in Zone 2 and Division 2 hazardous areas. It takes advantage of the relaxed design requirements for non-incendive (energy-limited) circuits compared with those for intrinsic safety. FNICO enjoys the same benefits as FISCO in terms of simple safety documentation and the elimination of cable parameter calculations, and retains the ability to connect and disconnect the field wiring in the hazardous area while under power and without 'gas clearance' procedures.

FNICO also has the following additional benefits compared with FISCO:

- Higher levels of bus current, allowing more devices to be connected to the hazardous area trunk.

- Easier selection of approved field devices. Suitable devices include EEx nL, non-incendive, IS (Entity) and IS (FISCO).
- Installation rules for non-incendive wiring are less onerous than those for intrinsic safety.

The requirements for FNICO systems are defined in IEC draft document 60079-27 Ed.1/CDV, and are expected to emerge in the final IEC 60079-27 standard, together with the requirements for FISCO.

The design requirements for FNICO power supplies, field devices, cable and terminators are essentially the same as those for FISCO, except in the following respects:

- The field wiring is classified as EEx nL in accordance with IEC 60079-15, or North American non-incendive.
- The maximum permissible value of I_o is 570mA.
- Field devices may have a maximum of 20 μH internal capacitance, thereby permitting IS "Entity" certified field devices to be used in FNICO installations.

4 – Documentation

All of the "pieces" of a Fieldbus system have been identified and it's now time to 'put it all together.' However, to do so, requires telling others what is wanted, how to connect it, how to size it, and how to represent this information on drawings. As the name implies, chapter 4 contains samples of how to do this, though not how to use the forms or worked examples.[1]

A Fieldbus device differs from a conventional instrument, only in the way it communicates with the remainder of the system. Therefore, the sensor specification will be the same as it is today, so it will be possible to continue using the ISA-20-1981 specification forms for this purpose. What will be different is the addition of a second page to capture the information needed for Fieldbus network design.

The key information to be derived from the "Fieldbus" data sheet includes:

- Function Blocks in the device and their associated execution time – this information is used for configuring and scheduling.
- Current requirements – needed for network design.
- Device capacitance – needed for network design.
- Network assignment – used for troubleshooting and maintenance so that once the data sheet is located, all other associated documentation can be found.
- Software/DD revision number – used for commissioning and maintenance.

1. Complete worked examples of these documents and drawings can be found in the author's book *Designing and Implementing a Foundation Fieldbus System.*

- Field devices containing a control algorithm shall be shown on the appropriate network drawing. Figure 4-1 shows a sample network drawing, including a field-based repeater and intrinsic safe barrier.

Figure 4-1 – Network diagram with repeater.

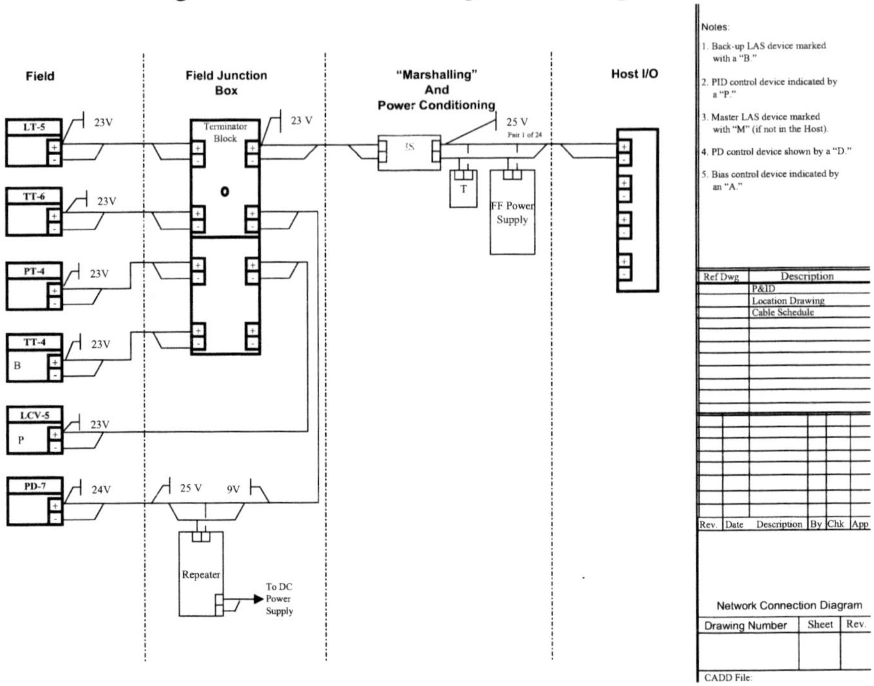

A complete list of all Fieldbus certified devices is always available and maintained by the Fieldbus Foundation at www.fieldbus.org.

Figures 4-2 and 4-3 show two different ways in which the information needed to specify the Fieldbus aspect of a device may be captured on a data sheet. Figure 4-2 assumes a single device per page while Figure 4-3 is more generic and is intended for multiple devices per page.

Figure 4-2 – Fieldbus data sheet: individual instrument.

Instrument Tag Number:

Basic Fieldbus Function Blocks	Segment Information
☐ Analog Input (AI) ___Number ___Execution Time (msec)	☐ Arithmetic (A) ___Execution Time (msec) ☐ Digital Alarm (DA) ___Execution Time (msec)
☐ Discrete Input (DI) ___Number ___Execution Time (msec)	☐ Calculate (C) ___Execution Time (msec) ☐ Analog Alarm (AA) ___Execution Time (msec)
☐ Bias/Gain Station (BG) ___Execution Time (msec)	☐ Deadtime (D) ___Execution Time (msec)
☐ Manual Loader (ML) ___Execution Time (msec)	☐ Complex Analog Output (CAO) ___Execution Time (msec)
☐ Proportional/Integral/ Derivative (PID) ___Execution Time (msec)	☐ Step Output PID (SOPID) ___Execution Time (msec)
☐ Analog Output (AO) ___Number ___Execution Time (msec)	☐ Set Point Ramp Generator (SRG) ___Execution Time (msec)
☐ Discrete Output (DO) ___Number ___Execution Time (msec)	☐ Signal Characterizer (SC) ___Execution Time (msec)
☐ Control Selector (CS) ___Execution Time (msec)	☐ Digital Human Interface (DHI) ___Execution Time (msec)
☐ Proportional/Derivative (PD) ___Execution Time (msec)	☐ ______________ ___Execution Time (msec)
☐ Ratio (R) ___Execution Time (msec)	**Device:**
Advanced Function Blocks	**Segment #:**
☐ Pulse Input (PI) ___Execution Time (msec)	**LAS Capable:** YES ☐ NO ☐
☐ Input Selector (IS) ___Execution Time (msec)	**Device current draw (mA):** **Device In-rush Current (mA):**
☐ Input Selector (IS) ___Execution Time (msec)	**Device Lift-off (Minimum)** **Voltage:** **Device capacitance:**
☐ Integrator (I) ___Execution Time (msec)	**Polarity Sensitive:** YES ☐ NO ☐
☐ Input Selector (IS) ___Execution Time (msec)	**Segment terminator location:** **VCR's:**
☐ Input Selector (IS) ___Execution Time (msec)	**DD Revision:**
☐ Input Selector (IS) ___Execution Time (msec)	**CFF Revision:** ITK Revision that Device was tested with:
☐ Signal Splitter (SS) ___Execution Time (msec)	**Notes** **Vendor to enter here all** **Non-Standard or Enhanced function block data:**
☐ Timer (T) ___Execution Time (msec)	**Vendor to enter here all unique Vendor Diagnostic/ Advance Diagnostics capabilities.**

Figure 4-3 – Fieldbus data sheet for multiple devices.

	1	2	3	4	5	6	7	8	9	10	11	12
TAG NUMBER												
Number of AI's												
AI Execution Time (msec)												
Number of AO's												
AO Execution Time (msec)												
Number of SS's												
SS Execution Time (msec)												
Number of TOT's												
TOT Execution Time (msec)												
Number of AR's												
AR Execution Time (msec)												
Number of PID's												
PID Execution Time (msec)												

Number of _____												
Execution Time												
Number of _____												
Execution Time												
Number of _____												
Execution Time												
Number of _____												
Execution Time												
Number of _____												
Execution Time												
Number of _____												
Execution Time												

Figure 4-3 – continued

Number of _____												
Execution Time												
Number of _____												
Execution Time												

Channel												
I.S. Segment (If Applicable)												
LAS Capable (Yes / No)												
DD Revision												
ITK Revision												
Polarity Sensitive (Yes / No)												
CFF Revision												
Notes:												

There is still discussion under way regarding whether Fieldbus should be shown on Piping and Instrumentation Diagrams (P&IDs) differently from conventional instrumentation. Figure 4-4 shows the symbology being considered to differentiate Fieldbus communications from other forms of serial communications.

Figure 4-4 – Digital communication signals.

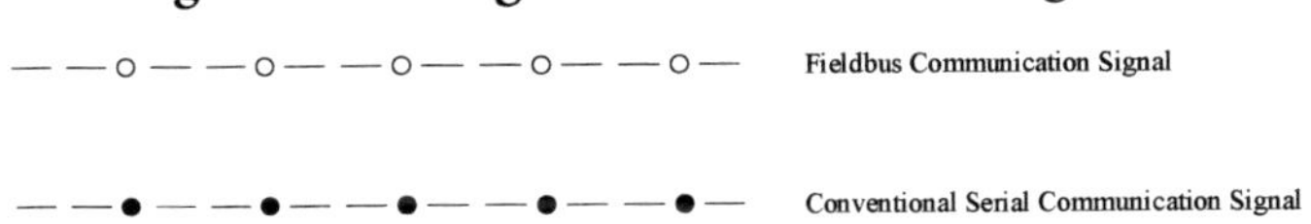

A number of other tables are listed on the following pages to help with the design and installation of a Fieldbus system.

Table 4-1 – System decision analysis.

Issue	Concern	Impact	Frequency	Combined Weight	Comment
		Weight	Weight		
1. Control in the Field					
			Sum		
2. Control in the Host					
			Sum		
3. Multi-variable Transmitters					
			Sum		
4. Etc.					
			Sum		

Other factors that may be worth considering include location of Power Supply, Galvanic versus Intrinsic Safe Isolation, etc. The weighting factors for the above are similar to those used in the Risk Analysis process and should be determined for each site. Consistent with the practices used in Risk Analysis, the "Impact" and "Frequency" analyses should be done independently of each other and then the overall weighting for each concern summed. The project leaders will then have a documented basis for selecting among various options for their project design.

The Network Decision Analysis is a two-stage process in that the team must agree on a Criticality Matrix and "hurdle rate" for the analysis proper. The following table illustrates how the Criticality Matrix may appear.

Table 4-2 – Network decision analysis.

	Criticality Weighting
System Trip (Criticality 1valve or similar)	10
Criticality 2 Final Control Element	
Criticality 3 Final Control Element	
Final Control Element with Control in Device	
Final Control Element without Control in Device	
Multivariable device, part of control loop, with LAS capability	
Multivariable device, part of control loop, without LAS capability	
Multivariable device, view only, with LAS capability	
Multivariable device, view only, without LAS capability	
Single variable device, part of control loop, with LAS capability	
Single variable device, part of control loop, without LAS capability	
Single variable device, view only, with LAS capability	
Single variable device, view only, without LAS capability	0

Weighting is assigned on the basis that 10 is most critical and 0 has no impact. No individual loop shall have an overall criticality rating of greater than *x** without specification deviation from the project owner/operator. (* *x* is to be determined by the project owner/operator.)

Table 4-3 – Device criticality decision matrix.

Device Tag	Criticality Level	Criticality Weighting		Loop Number	Function Block Execution	Current Consumption (mA)	Inrush Current (mA)
			Sum		Time		

The above table also captures the critical information required for design of the balance of the Fieldbus network on the right-hand side of the double border.

Table 4-4 – Connector decision analysis.

Issue	Concern	Impact		Frequency		Combined Weight	Comment
			Weight		Weight		
1. Screw Terminals							
					Sum		
2. Quick Connections							
					Sum		
3. Spur short circuit protectors							
					Sum		
4. Etc.							
					Sum		

The project site may already have a standard termination option they prefer. If not, this table and process, which are similar to those of the System Decision Analysis table, can be used to make an informed decision.

Segment Loading Calculation

There are two different forms of loading calculation: one for the IS or NIS installation and another for projects using the FISCO installation method.

Table 4-5 – FISCO installation.

Trunk			Spur Cable		
		Okay			Okay
R'			R'		
L'			L'		
C'			C'		
L_{max}			I_{sup}		
Power Supply			**Category**		
V_{PS}			U_o		
I_{PS} (Amps)			I_k (Amps)		
Power Supply Location	Host or Field JB		P_{max} (Watts)		
Spur protection short circuit load (mA)			Output Characteristic	Trapezoid, Rectangular	

Device					Segment
Tag	C_D	I_D (mA)	ΣI_D	Okay	ΣC_T

Table 4-6 – IS/NIS installation.

Trunk			Spur Cable			Power Supply	
R		Ω/km	R		Ω/km	V_{PS}	
C		nF/km	C		nF/km	I_{PS} (mA)	
L		Meters				Power Supply Location	Host or Field JB
Voltage Drop:		Volts	L_{TMAX}		meters	Spur protection short circuit load (mA)	

Device						Check		Segment		
Tag	V	I (mA)	C (nF)	Spur (m)	V_d	$V_d > V_{min}$	C_S	C_T	I_N	A

Below is a partial checklist that can be used for configuring a Fieldbus system. Depending on the nature of the installation, additional items may need to be added to this list, which is intended to provide an indication of the types of items requiring consideration.

Table 4-7 – Configuration worksheet/checklist.

	Parameter	Complete	Initials
Device	Tag match		
	DD Revision		
	Transducer Block Mode		
	Resource Block Mode		
	Function Blocks and mode		
Network	Cyclic/Acyclic time ratio		
	Schedule published/downloaded		

5 – System Integration

Everything is now built. Time to configure the system so all the parts work together. Chapter 5 discusses configuration and scheduling, or who "talks" directly to who and when.

When a message is transferred, it goes down through a channel called Virtual Communication Relationship (VCR) to add Physical Device (PD) before it goes to the wire. At the destination, it goes up through the partner VCR to the receiving application. Process-control information packets are appended and removed when a message goes through VCRs to allow layers to perform their specific functionality. Figure 5-1 represents how data is transmitted through the layers and across the Data Link Layer via VCRs.

Figure 5-1 – Fieldbus VCR communications.

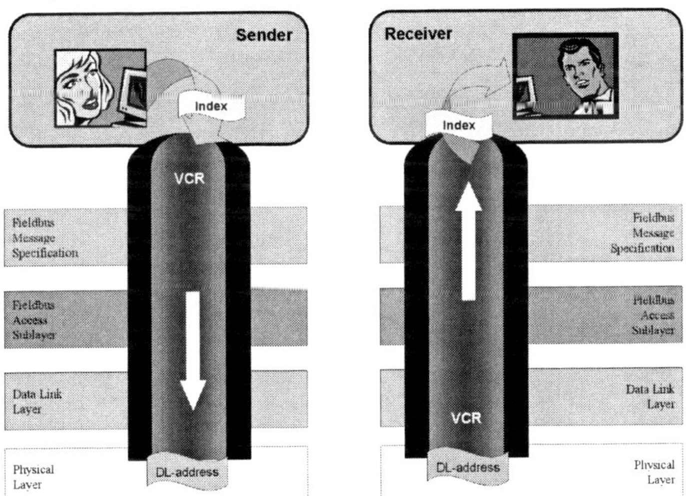

Before starting configuration, the following files will be required:

- Device commissioning files–Device Description files .sym and *.ffo
- Network configuration file–Capabilities file *.cff

Capabilities File is a common text format file that can be read and interpreted by a host so that it can create the necessary internal files and database links it requires to use the Device Description files.

The Fieldbus Foundation Web site maintains the most recent version of all these files for every certified device. It is important to manage the revisions of these files and hence the software version of the devices themselves since they are not always compatible between different revisions, especially if the host and field device are at different revision levels.

5.1 Configuration

Network Configuration is responsible for giving the correct information of the index and Data Link (DL) address, as well as other operating information, to VCRs through Network Management.

There are different types of VCRs: Client–Server, Report (Sink–Source) Distribution, and Publisher–Subscriber.

Client–Server VCR Type

The Client–Server VCR type is used for queued, unscheduled, user-initiated, and one-to-one communication between devices on the Fieldbus.

Queued means that messages are sent and received in the order submitted for transmission, according to their priority, without overwriting previous messages.

The Client–Server VCR type is used for operator-initiated requests such as setpoint changes, tuning parameter access and change, alarm acknowledge, and device upload and download.

Report Distribution VCR Type

The Report Distribution VCR type is used for queued, unscheduled, or user-initiated one-to-many communications.

Tip 12 – Fieldbus devices that send alarm notifications to the operator consoles typically use the Report Distribution VCR type.

Publisher–Subscriber VCR Type

The Publisher–Subscriber VCR type is used for buffered, one-to-many communications.

Buffered means that only the latest version of the data is maintained within the network. New data completely overwrites previous data.

The Publisher–Subscriber VCR type is used by the field devices for cyclic, scheduled publishing of User Application Function Block input and outputs such as process variable (PV) and primary output (OUT) on the Fieldbus.

Table 5-1 summarizes the different VCR types and their uses.

Table 5-1 – VCR types and their uses.

Client–Server VCR type	Report Distribution VCR type	Publisher–Subscriber VCR type
Used for operator messages	Used for event notification and trend reports	Used for publishing data
Setpoint changes Mode changes Tuning changes Upload/download Alarm management Access display views Remote diagnostics	Send process alarms to operator consoles. Send trend reports to data historians.	Send transmitter PV to PID Control Block and operator console.

Each device requires a minimum number of VCRs to communicate with each other and with a host. Below are suggested minimum guidelines to be used when selecting devices on a segment.

Each device requires the following five basic Device Blocks:

1. One Client–Server for each Management Information Base (MIB)
2. One Client–Server for the primary host
3. One Client–Server for the secondary host or maintenance tool
4. One Report Distribution for Alerts
5. One Report Distribution for Trends

Tip 13 – Not all hosts support trends and alerts so these two blocks may not be used, though they must be included in the calculations in case this capability, as specified in the Fieldbus documents, is added in the future.

Each Function Block requires the following:

- One Publish–Subscriber for each I/O.
- If the Function Block is used internally, it does not need a VCR.
- An input parameter may be linked to another input parameter but only in the same device.

Depending on the type of actuator being used, action can be set in both the PID and AO Blocks, so there is a risk of inconsistency and confusion if 100% sometimes means open and other times means closed. Therefore, it is best to be consistent and implement a control strategy. For example, a PID output of 100% means a valve is fully open, independent of the actuator being air to open or air to close. This can be done by configuring the AO Block so that the I/O option is "Increase to close" for air-to-close actuators.

As stated elsewhere, Fieldbus has various levels of alarms, which allows interlocks to be suited to the process. For example, for an important shutdown interlock, action should be taken on either BAD or UNCERTAIN data quality. Lost communication is always indicated as BAD status and resulting loop shutdown.

The result is that to improve system safety, the status should not only be displayed to the operator but should also be included as part of the system interlock logic.

Internal Fieldbus and device diagnostics are far better at identifying errors than identifying external discrepancy logic. For high availability, which often contradicts high safety, the loop should be configured to shut down only when the status is BAD; an UNCERTAIN status will only alert the operator and, if appropriate, maintenance personnel.

Table 5-2 from Berge's *Fieldbus for Process Control* summarizes suggestions on how to configure a network for safety versus availability.

Table 5-2 – Configuring a network for safety vs. availability.

Parameter	Safety	Availability
STATUS_OPTS	Don't (i.e., as BAD)	Use UNCERTAIN as GOOD
STATUS_OPTS	Target to Manual if BAD IN	Don't (i.e., return to normal at once)
STATUS_OPTS	Initiate fault state if IN is BAD	Don't (go to manual)
STATUS_OPTS	Initiate fault state if CAS_IN is BAD	Don't (go to automatic)
STATUS_OPTS	Set Target mode to manual if IN is BAD	Don't (return to normal once okay)
STATUS_OPTS	Set quality as UNCERTAIN if limited	Don't (consider GOOD)
STATUS_OPTS	Set quality as BAD if limited	Don't (consider GOOD)
STATUS_OPTS	Set quality as UNCERTAIN if block is manual mode	Don't (consider GOOD)
IO_OPTS	Fault state to value	Don't (freeze output)
IO_OPTS	Use Fault State value on restart	Don't (restart from present position if available)
Function block link state count limit	Few	Many
SHED_OPTS	No return	Normal return
FSAFE_TIME	Short	Long
FEATURE_SEL	Fault state supported	Don't (disabled)
Backup LAS Function	Disabled	Enabled
SHED_RCAS	Short	Long
SHED_ROUT	Short	Long

Tip 14 – A Backup Link Active Scheduler (LAS) shall be configured for all control loops and should normally reside in the device with the minimal processing load, for exam-

ple, a temperature transmitter. Again, this is so the loop can continue to operate, in a controlled manner, as long as it has power.

Systems shall be configured for control bumpless "fail over" to the Backup LAS if control or communication is lost with the Primary LAS, which is often in the host control system.

Where the transmitter and control valve in a loop cannot be conveniently wired to the same segment, control shall normally reside in the host controller. Because the network communications must go through the host control it is normally done there to minimize the demands on the field devices.

For cascade loops the primary measurement should reside on a separate H1 card from the secondary measurement since this reduces the potential for a common mode failure to affect loop control.

Fieldbus devices provide support for Trend Function and Alert Blocks to reduce segment traffic by transmitting multiple signals as one block.

A shorted segment or segment power supply failure shall send valves to their failure position, regardless of the device hosting the PID algorithm.

"Fail over" Strategies and Design Considerations

The control narrative and project risk assessment(s) will determine how each segment and device should operate under normal operating and failure conditions. With Fieldbus, it is possible to specify different failure strategies depending on the type of failure. For instance:

- Loss of Communication

- with host–continue operations with control in the field, fail open, fail closed, or fail last.
- with segment–fail open, fail closed, or fail last.

- Loss of Power–fail open, fail closed, or fail last.
- Bad AI Signal–fail open, fail closed, or fail last.
- Cascade Failure–PID Control only, fail open, fail closed, or fail last.
- Loss of motive fluid (instrument air)–fail open, fail closed, or fail last.
- Etc.

Provided power is still available on the network, automatic switchover of the process automation network shall not interrupt other system operations. Automatic switchover from the process automation network to the individual segment shall not interrupt other segment devices such as Backup PID or Backup LAS.

Scenarios where the loop would go to its fail position would be a broken segment, shorted H1 segment, failed H1 power supply, or failed H1 filter–isolator.

A system can be configured for control class blocks to use the status option USE UNCERTAIN as GOOD so they will treat UNCERTAIN as either GOOD or BAD. The default value is to treat UNCERTAIN as BAD.

The process variable filter time parameter, PV_FTIME, is used to set the time constant (63% of steady-state value) in seconds of a first order lag. If the time is set to zero seconds, the damping is disabled.

Figure 5-2 is indicative of the final result of a configuration screen on a typical host.

Table 5-3 shows the relative priority assigned to each of the Fieldbus operating modes. A device in Out Of Service (OOS) mode will cause the loop to move to this state even if the other components of the network remain in automatic mode.

Figure 5-2 – Host configuration screen.

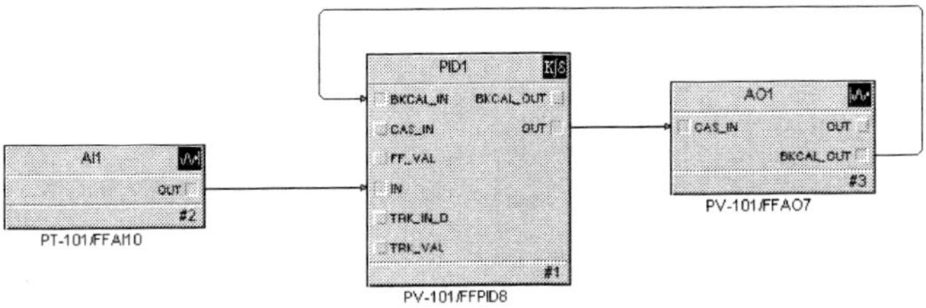

Table 5-3 – Fieldbus operating mode priorities.

Mode	Description	Setpoint source	Output source	Priority
OOS	Out of Service	Operation (SP)	Operator (OUT)	7 Highest
IMan	Initialization Manual	Operator (SP)	Lower Block (BKCAL_IN)	6
LO	Local Override	Operator (SP)	Other Block (TRK_VAL)	5 Control Class
		Safe Value (FSAFE_VAL)		Output Class
Man	Manual	Operator (SP)	Operator (OUT)	4
Auto	Automatic	Operator (SP)	This block	3
Cas	Cascade	Higher Block (CAS_IN)	This block	2
RCas	Remote Cascade	Other Application (RCAS_IN)	This block	1
ROut	Remote Output	Operator (SP)	Other Application (ROUT_IN)	0 Lowest Priority

Fieldbus has 15 levels of alarms, as well as several orders of magnitude more information than is available with a conventional DCS or analog control system. Configuration must not only handle and assign actions to each of these 15 alarm levels, it must also route the various signals to the correct output device or system. Table 5-4 summarizes the alarm levels of a Fieldbus system.

When devices or communications fail, the associated control loops cannot continue to operate. Fieldbus has a "Fault State" function to monitor the status of every loop. When the Fault State is activated, the output class block goes into local override mode. However, for the Fault State function to work, it must be enabled by using the "Fault State Supported" option in the Feature Selection (FEATURE_SEL) parameter in the Resource Block. The Fault State function can be activated in one of four ways:

Table 5-4 – Fieldbus alarm levels.

Alarm	FF Alert Levels
15	Critical
14	Critical
13	Critical
12	Critical
11	Critical
10	Critical
9	Critical
8	Critical
7	Advisory
6	Advisory
5	Advisory
4	Advisory
3	Advisory
2	Low – fixed
1	No notification
0	No indication

1. "Initiate Fault State" on the cascade setpoint input

2. Cascade setpoint input communication watchdog times out

3. Power on

4. Forced from the Resource Block

As with devices, the HMI of a Fieldbus system will not be significantly different from the one the process operators see today. The major enhancement that is possible with Fieldbus is that signal and loop status can be communicated to them in real time. As a result, the panel operators will have more confidence in the control system since they will always know how reliable any given signal is.

5.2 Scheduling

As with any network, the more information transmitted in a given time frame, the faster the network and all its components must operate to keep pace with this information. Since Fieldbus H1 is constrained to a rate of 31.25 Kbps, the only way to transmit additional information is to increase the cycle time.

A link can carry about 30 scheduled messages per second. This means the network could have 3 devices, each sending 10 messages per second, or 120 devices, connected by repeaters, each sending 1 message every 4 seconds.

Tip 15 – A good rule of thumb to use for the initial estimate related to bandwidth requirements is to assume each device requires 50 ms to execute its Function Block.

Therefore, the total bandwidth required can be estimated using the following formula:

$$t_{LOAD} = (N_P + N_C) \times 50\,\text{ms}$$

where:

t_{Load}	=	time to execute all Function Blocks in loop
N_P	=	number of Publishers (devices on the network)
N_C	=	number of communications with the HMI

The minimum unscheduled/acyclic time shall be 70–80% for a newly commissioned segment. This allows room for future growth, if required.

For example, assume a 1-s macro LAS cycle (macrocycle) gives 150 mS for scheduled data transmission, with 70% of the available 500 mS available for acyclic communications. The time available for future use in this case would be 350 mS.

Cyclic traffic time may be determined by the summation of the individual Function Block execution times plus the publish time on the network.

Sum the run time of each device's Function Blocks. For our example, an AI Block runs in 50 ms, a PID Block in 150 ms, and an AO Block in 100 ms; and:

Tip 16 – As a rule of thumb, each external link (through the Fieldbus channel) uses about 25 ms.

For the loop shown in Figure 5-3 in which the PID and AO Blocks are located at the valve's positioner, the macrocycle is estimated as 325 ms.

Figure 5-3 – Loop configuration.

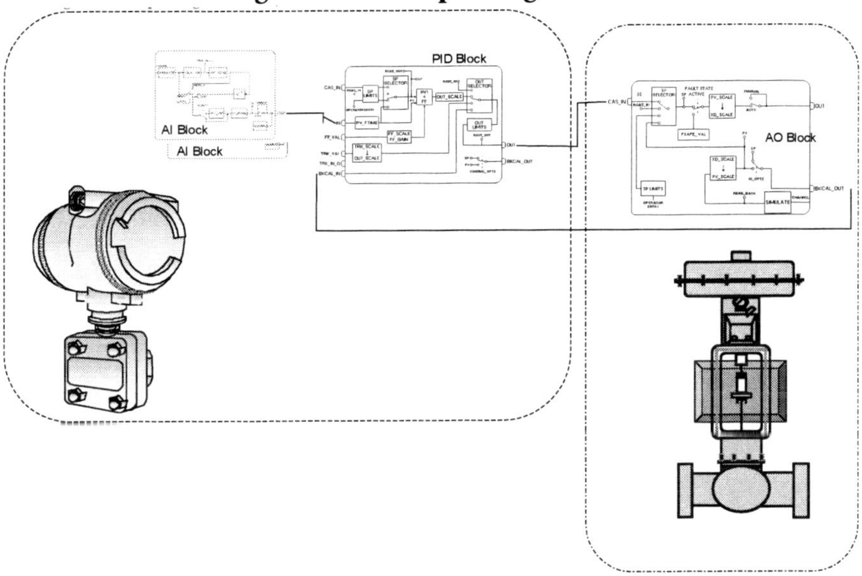

If a second configuration identical to this one were on the same network, the macrocycle would not be the double because as the blocks run in parallel, the new macrocycle time would be 350 ms. This is because only the time corresponding to the extra link included in the network must be added.

As a general guideline in the hydrocarbon industry, the number of devices for segment execution time should be as follows:

- For loops requiring 1-s execution time, limit segment to eight devices with a maximum of three valves (four valves with client approval where all loops are simple loops with control in the valve positioner).

- For loops requiring 0.5-s execution time, limit segment to four to eight devices.
- For loops requiring 0.25-s execution time, limit segment to fewer than three devices with a maximum of one valve.

The above are suggested strategies to manage the risk associated with a single component failure in a Fieldbus system. Since each and every facility has a different risk management strategy and levels of "acceptable risk" they can each economically manage, the project team needs to determine a set of guidelines such as these at the start of their project.

Tip 17 – Do not mix devices with different macrocycle times (1 s versus 0.25 s) on the same segment. Mixing of macrocycles can lead to schedules that may not be within the capability of some Link Masters.

Mixing macrocycles requires diligent design practices with particular attention to the possibility of "periodicity" and hence conflicts after a significant number of cycles. A problem of this type will be difficult to diagnose and hence the recommendation to "keep it simple."

The cardinal rule for scheduling networks is that they must have a minimum amount of acyclic time in each cycle *and they must operate at a higher frequency than the process.* The acyclic time is required to communicate information other than the Function Block parameters, including alarms and configuration information, while the cycle proper must be sufficiently fast to ensure that it is representative of the process changes. The minimum recommendation is that the network cycle time be three times that of the process, though a sampling frequency of six is preferred.

If the residence time of an inlet separator is 2 min, then the minimum cycle time for loop LIC-1 is 120/3 = 40 s.

Tip 18 – A good rule of thumb to start a design is to use the typical cycle time for a traditional control system, which is 2 s per I/O scan.

If a control loop must communicate through the host because AO devices such as a variable-speed drive pump are on a different control network, the loop response time will have to be calculated as not only the response time of the H1 network, but also the time it takes to have (1) the H1 network scanned by the host, (2) the host to act on this information, (3) another cycle to complete, and then (4) at least two more cycles to complete in the other network, such as the ControlNet system, for the change to occur at the pump.

Figure 5-4 shows a possible format for calculating the cycle time and number of VCRs required for a Fieldbus network.

Figure 5-4 – Segment bandwidth calculation.

Network: ______________ Target Macrocycle (sec): ___________ % Acyclic Time: ___

Host Limitations					
Network			System		
Maximum Number of Function Blocks:			Maximum Number of Function Blocks:		
Min. Macrocycle time:		msec	Min. Macrocycle time:		msec
Target Macrocycle time:		sec	% Acyclic Time		

Tag	Function Block	Cycle Time (msec)	Publish	Subscribe	Internal	Cum. Cycle Time	Check

AI–PID–AO Timing (Host Versus Field Execution Strategies)

Everyone is familiar with the fact that the PID algorithm requires the "error" to work, and this text has also shown how VCRs are used for publishing/subscribing (reading) data. This explains why placement of the PID Block can affect timing. Let's walk through some examples.

1. PID in AO device
    - AI publishes its process value and the AO/PID subscribe (read).
    - PID performs its calculations and sends its output internally to the AO block, which adjusts and replies with the BACK_CALC, then publishes its value to complete the cycle.
2. PID in host
    - AI publishes its process value and host subscribes.
    - PID is calculated and published for the AO to subscribe.
    - AO moves and publishes for PID to subscribe and read the BACK_CALC.
    - Cycle completed.
3. PID in AI device
    - AI internal communication to PID Block. PID Block determines the OUT value. Both the Out and AI value are published to the network.
    - AO subscribes to PID_OUT and makes it move, then publishes the new AO-OUT value.
    - Cycle completed.

As can be seen by the number of steps representing Publisher–Subscriber messages on the network, keeping the PID in the host does have an impact on the amount of time it takes for a PID algorithm to be completed.

6 – Commissioning

This is it, time to turn things on and see them work as a Digital Control System. Chapter 6 presents a number of procedures, tips, and tricks to help make this process as smooth as possible.

Make sure that all final checks have been completed before you start up your measuring point.

Tip 19 – The technical function data of the FOUNDATION™ Fieldbus interface to IEC 61158-2 must be maintained.

The bus voltage of 9–32 V and the current consumption at the measuring device can be checked using a normal multimeter.

The network has the correct number of terminators–only one at each end.

Cable integrity and signal strength can be tested with a number of handheld meters from Relcom Inc. These meters are shown in Figure 6-1.

The device must have a default node address in the range 248–251. The Fieldbus system will assign the device a final operational address in the range 16–247.

The files required for commissioning and network configuration can be obtained from the manufacturer or at the Fieldbus Foundation Web site (www.fieldbus.org).

Figure 6-1 – H1 network analysis tools.

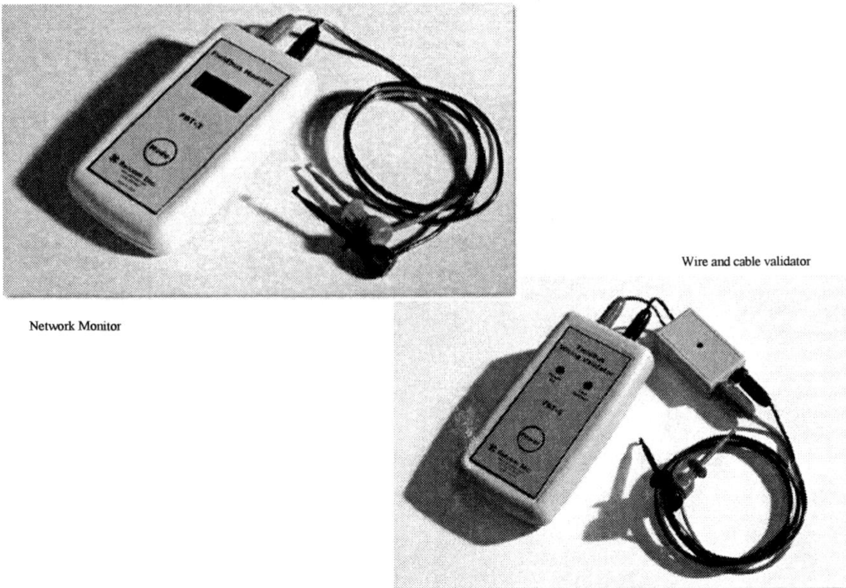

Network Monitor

Wire and cable validator

The device is identified by the FOUNDATION™ Fieldbus in the host or configuration system via the device ID (DEVICE_ID). The DEVICE_ID is a combination of the manufacturer ID, device type, and device serial number. It is unique and can never be duplicated.

The following description allows step-by-step commissioning of the measuring device and all the necessary configuration for the FOUNDATION™ Fieldbus:

1. Switch on the Measuring Device.

   Note the DEVICE_ID on the device nameplate.

2. Open the configuration program.

3. Load the device description file or CFF file into the host system or into the configuration program. The first time it is connected, the device will report something similar to the following:

Display text (XXX = serial number)	Description
DEVICE_NAME xxxxxx	Field device tag name (PD-TAG)
nnnannnnn-	Device ID
RESOURCE_xxxxxxxxx	Block name - Resource Block
TRANSDUCER_ xxxxxxxxx	Block name - Transducer Block
ANALOG INPUT_1_xxxxxxxxxxx	Block name - Analog Input Function Block 1
PID_xxxxxxxxxxx	Block name - PID Block

Tip 20 – Be sure the device is supplied with a bus address that is in the address range reserved for re-addressing of field devices, between 248 and 251. This means that the LAS (Link Active Scheduler) automatically assigns the device a free bus address in the initialization phase.

Tip 21 – Each DEVICE_NAME in the system must be unique and can be up to 32 characters in length. It is recommended that suffixes be used to identify the signal and block type.

Table 6-1 provides examples of how this may be applied.

Table 6-1 – Examples of identifying signal and block type.

Analog Input signal 1	PT-1_AI1
Analog Input Signal 2	PT-1_AI2
Transducer Block	PT-1_TB
Resource Block	PT-1_RB
PID Block	PT-1_PID

4. Identify the field device using the DEVICE_ID that you noted down and assign the desired field device tag name (PD_TAG) to the Fieldbus device in question.

Configuration of the Resource Block

5. Open the Resource Block.

6. On delivery, write protection is disabled so that you can access all the write parameters. Check this status via the parameter WRITE_LOCK:

   - Write protection activated = LOCKED
   - Write protection deactivated = NOT LOCKED
   - Deactivate the write protection if necessary.

7. Enter the desired block name.

   Factory setting: RESOURCE_ xxxxxxxxxxx

8. Set the operating mode in the parameter group MODE_BLK (parameter TARGET) to AUTO.

Configuration of the Transducer Block

9. Enter the desired block name. Factory setting: TRANSDUCER_ xxxxxxxxxxx

10. Open the Transducer Block.

11. Set the operating mode in the parameter group MODE_BLK (parameter TARGET) to OOS, that is, block Out Of Service.

12. Now configure all the device-specific parameters for the measurement.

Tip 22 – Changes to the device parameters can be made only after entering a valid access code in the Access Code parameter.

13. Set the operating mode in the parameter group MODE_BLK (parameter TARGET) to AUTO.

Configuration of the Analog Input Function Block

Some Fieldbus devices have multiple Analog Input Function Blocks that can be assigned optionally to different process variables. The following description provides an example for Analog Input Function Block 1.

14. Enter the desired name for the Analog Input Function Block. Factory setting: ANALOG_INPUT_1_ xxxxxxxxxxx
15. Open the Analog Input Function Block.
16. Set the operating mode in the parameter group MODE_BLK (parameter TARGET) to OOS, that is, block Out Of Service.
17. Using the parameter CHANNEL, select the process variable that is to be used as input variable for the Function Block algorithm (scaling and limit value monitoring functions).
18. In parameter group XD_SCALE, select the desired unit of measure and the block input range (measurement range of the flow application) for the process variable in question.

Caution: *Make sure that the selected unit of measure is suitable for the measurement variable of the selected process variable. Otherwise the parameter BLOCK_ERROR will display the error message "Block Configuration Error".*

19. In the parameter L_TYPE, select the type of linearization for the input value (Direct, Indirect, Indirect Sq Root).

Caution: *Note that with the type of linearization "Direct," the configuration of the parameter group OUT_SCALE must agree with the configuration of the parameter group XD_SCALE. Otherwise the block-operating mode cannot be set to AUTO. Such incorrect configuration is indicated via the parameter BLOCK_ERROR (Block Configuration Error).*

20. Use the following parameters to define the limit values for alarm and warning messages:

HI_HI_LIM	limit value for the upper alarm
HI_LIM	limit value for the upper warning
LO_LIM	limit value for the lower warning
LO_LO_LIM	limit value for the lower alarm

The limit values entered must be within the value range specified in the parameter group OUT_SCALE.

21. In addition to the actual limit values, you must also specify the action taken if a limit value is exceeded using so-called "alarm priorities" (parameters HI_HI_PRI HI_PRI LO_PR LO_LO_PRI).

Tip 23 – Reporting to the Fieldbus host system takes place only if the alarm priority is higher than 2.

22. System configuration/connection of Function Blocks as shown earlier in Figure 5-2.

A concluding "overall system configuration" is essential so that the operating mode of the Analog Input Function Block can be set to AUTO and so that the field device is integrated into the system application.

To do this, configuration software is used to connect the Function Blocks to the desired control strategy–generally graphically–and then the sequence of the individual process-control functions is specified.

23. After specifying the active LAS, download all the data and parameters into the field device.
24. Set the operating mode in the parameter group MODE_BLK (parameter TARGET) to AUTO. However, this is possible only if two conditions are met:

- The Function Blocks are correctly connected with each other.
- The Resource Block is in operating mode AUTO.

7 – Troubleshooting

Congratulations, you are up and running. Time now to "fine tune" and optimize this system. This chapter offers a few things to consider and tools available to do just that, to keep a Fieldbus system in "top shape."

The quality GOOD (NONCASCADE) means the value may still be used for control, despite the fact that signal quality is deteriorating.

Tip 24 – The stale rate on PID Blocks must be set to a minimum of 3.

If the PID stale rate is left at the default state of 1, the loop can periodically "cycle" between in and out of service if there is a sporadic communication issue on the segment, which causes the AI Block to be missed. A side effect of cycling in and out of operation is inadvertent changing of setpoints when setpoint tracking is specified.

One of the few tools available to analyze a Fieldbus network is Relcom Inc.'s FBT-3 Network Monitor. This device uses a simple interface to check key traffic and signal level characteristics, which can be checked against the enclosed manual.

Another network analysis tool, though much more complex, is the Bus Monitor software from National Instruments. This combination hardware and software package resides on a laptop or personal computer connected to the network as a guest and is capable of monitoring all aspects of network traffic and packets.

Tip 25 – The National Instruments equipment can be configured as a Host or Guest. If it is configured as a Host, it will conflict with the already existing Host and likely cause the entire network and/or system to crash.

When the output from the secondary control block of a control cascade reaches the limit for the setpoint in the primary control block, the primary will set limited status in its back calculation output to the secondary so as to prevent the upper block from moving its output further in that direction. The secondary will not change mode.

If the LAS fails, the Backup LAS assumes control of the network. The result is that the loops may be running on outdated information of a few seconds. To prevent this from accelerating to cause a problem, limits should be placed on the time a loop can operate in this alternating LAS primary node before having the loops "gracefully" shut down.

7.1 Tuning

Figure 1-4c represents the "internal" functions that reside in the PID Function Block. To make the tuning constants for the PID Block dimensionless, the PV_SCALE parameter must be configured.

Three tuning parameters are in a PID loop: Proportional Gain (GAIN), Integral Reset Time (RESET) in seconds per repetition, and Derivative Time Constant (RATE) in seconds. To disable the integral function, set RESET to +Inf. Derivative action (RATE) is set to zero to disable this action.

Caution: *Many different forms of the PID algorithm are implemented in the various PID Blocks of different manufacturers. To ensure bumpless transfer of a loop, be sure of the form implemented in the PID Blocks of the loop.*

The CONTROL_OPTS parameter can be used to configure the output as direct or reverse action with the default being reverse action.

Caution: *When the controlled variable input to a primary PID block in a cascade loop is "bad," the primary's PID algorithm cannot function. It is possible to let the secondary PID control analog output block pass directly to the primary PID. Caution is necessary since not all control schemes are stable in bypass. To minimize the risk, enable the bypass feature by setting the CONTROL_OPTS parameter option to "Bypass Enable." Bypass itself is activated with the BYPASS parameter, which can only be set in manual or OOS mode.*

8 – Operations & Maintenance

Fieldbus introduces a number of changes to the traditional ways of doing business in a processing facility. Fully deploying and using Fieldbus technology in a facility, with associated changes in operating procedures can result in savings of up to 50% of the automation maintenance budget alone. This chapter highlights just a few of the changes required to be able to realize the benefits of a fully digital control system.

8.1 Operations

Digital communications technology provides operations the potential for several orders of magnitude more operational data than is available from a traditional analog field device/system. Key among the additional information is the "status" signal sent along with every process variable transmission.

This status information can be used to inform the panel operator that despite what they think, the device is in fact reading correctly. In addition, rather than have the maintenance technician reference or calibrate the device, the status information should instead be included in the decision-making process. Similarly, application engineers can make use of the status information in their control algorithms/applications so that they are only run when the signals on which they rely are known to be operating properly.

Since Fieldbus has multiple levels of "failure" and alarming it is also possible to configure the control system to respond in different ways depending on the nature of the status being broadcast. Obviously, if the device status is "bad" or "OOS – Out Of Service" meaning it has failed; the signal should not be used for control. If however the device status is "Uncertain," the decision is not as clear and needs to be determined by the control and process staff as to how this could impact the operation and then have the control system respond accordingly.

The key to successfully implementing Fieldbus is to only send information to the people that they can use to improve the reliability and profitability of your operation. Doing otherwise is counterproductive and results in "information overload" where the important information the operator requires cannot be found among all the other data with which the process operator is being bombarded.

8.2 Maintenance

Maintenance departments are the people in a facility to be most impacted by the introduction of Fieldbus technology, not only because of the technology itself but also because of the associated changes that will result in traditional work practices. The technicians will also have to learn how to work with a number of new tools, most important of which will be a laptop or other computer with associated specialized software for in-depth system analysis.

The two most significant changes with Fieldbus versus analog control systems are:

1. Fieldbus devices MUST be connected to a network for them to be "live" and worked on for such things as calibrating[1]/referenc-

---

1. Modern digital transmitters are HIGHLY accurate devices for which a field calibration will not be nearly as accurate as what can be achieved in a factory setting. Some manufacturers are quoting stability guarantees of 10 years on their devices at levels better than were possible with earlier generations of equipment. Therefore the term calibrate should be replaced with the term "reference" or "reference check" as the bench test is simply to confirm that the device is configured for the range specified on the data sheet.

ing, re-ranging or any other work that needs to be performed on a "live" device.

2. Fieldbus devices are part of the control system network and therefore any changes made to the device are 'propagated' through the system and will likely change the configuration in the DCS once they have been connected. If they do not change the host system configuration, they will as a minimum raise an alarm so change management procedures need to be established for all work done on the host or field devices, including who (application engineer or maintenance technician) is responsible for what.

Because the sensing technology on which Fieldbus devices are based is unchanged from the past, though it is now more accurate, the majority of new problems with Fieldbus systems arise from something in the Physical Layer. Maintenance technicians must therefore learn how to gather information on the Physical Layer so they can determine the most likely cause of the network communication problems.

Fortunately, the traditional multimeter is still useful for measuring the voltage at each juncture of the network:

- Power Conditioner inlet and outlet
- Field Junction boxes
- Field Devices

The multimeter should also be able to measure the capacitance of the segment since this parameter also has an impact on the signal communication losses.

A complete set of these measurements should be taken as soon as the field cabling is installed as a baseline measurement and then again once the entire segment is commissioned and lastly after every change (addition or removal) of a device on the segment. Doing this will identify grounding problems and provide a baseline to work from in the future if problems should occur.

Relcom Inc. manufactures a useful tool, the FBT-3 and now also the FBT-6 that "polls" key physical layer parameters:

Volts – DC voltage on the network at the point the device is connected

LAS – Signal level of the Probe Node should be greater than 150 millivolts

Device – Number of devices on the segment as well as an indication if any changes in the number of devices since the last cycle.

Noise Av – Average noise on the network. This should be less than 75 millivolts.

Noise Pk – Peak noise recorded since the device has been connected to the network.

Invert – Hexadecimal address of any device transmitting an inverted Fieldbus signal.

FEC – Framing Error Code detects if the tester starts to decode a frame and is unable to finish.

Other manufacturers make similar devices and new product is regularly being released that may have similar or additional functionality to that described above. The Relcom device is one that is widely deployed in the industry to provide a preliminary inspection of the health of a Fieldbus network.

These data points are useful in determining the possible cause of problems and serve as a starting point to work from with the more sophisticated tools available on a laptop or other computer.

Many new power conditioners support the transmission of their operational status to the control system either via an Ethernet or other connection or as a minimum with a solid state contact to raise an alarm that potential problems with the Physical Layer are or soon will be happening.

Most important to the lifecycle operation of a facility is the diagnostic data available from Fieldbus devices when connected to a computerized maintenance system. This data makes it possible to do prognostic maintenance of the control system, with the associated possibility of reducing maintenance costs versus a traditional analog only system by anywhere from 25% to more than 50%.

9 – New Developments

9.1 Fieldbus Safety

At the time of this second edition, the Fieldbus Foundation is developing its first Fieldbus safety products for expected release in 2006. Devices that receive the Fieldbus "check mark" as complying with the safety standard indicates that the device is compliant with the new standard developed to be consistent with IEC 61508 and verified by TÜV. The "check mark" does *not* mean, however, that the device itself is safety certified to a certain SIL level. The Fieldbus specification has been designed so that the devices will meet a SIL 2 rating. The actual SIL rating for the device will continue to be the responsibility of the manufacturer and associated safety certifying organizations to have the device itself certified to the appropriate SIL rating.

The Fieldbus safety bus, as with most other safety buses, uses a "black channel" model as its basis. This means that rather than develop a new communications protocol from scratch, the safety protocol adds a number of protections and other features to ensure the timing/transmission, as well as the integrity of communications between devices. Consequently a number of new standards needed to be developed and most of the existing standards needed to be updated to include the required new functionality. In addition, a number of new safety specific function blocks have been defined to allow for safety logic control in the field, similar to that made possible for conventional devices.

Figure 9-1 shows how the IEC 61508 requirements have been added to either end of the H1 communications channel.

Figure 9-1 – SIS user layer extensions.

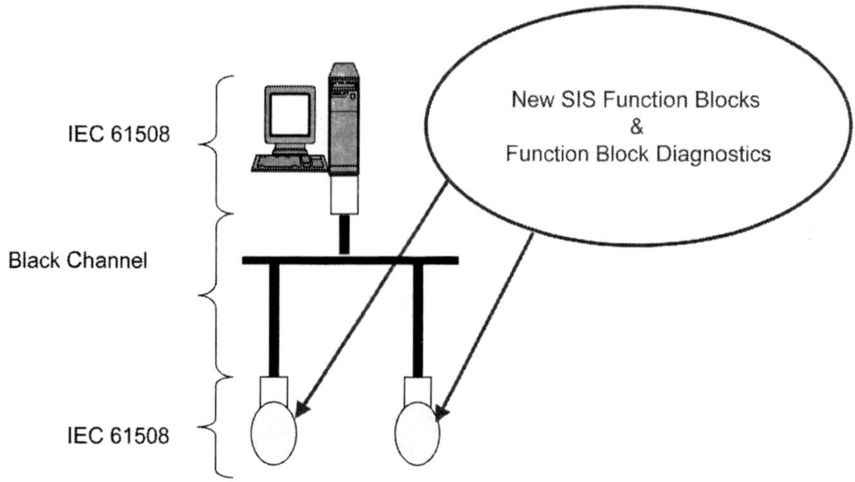

One of the other major advantages of using a "black channel" model is that the physical layer for the safety network will be the same as the "standard" protocol, and if desired, a single network could contain both safety and non-safety-related devices.

Figure 9-2 shows how a 2oo3 (2 out of 3) voting scheme could be implemented in the field using the Fieldbus safety bus. The individual physical devices are shown inside the rectangles, and like control in the field, they must all reside on the same network.

Figure 9-2 – Example SIS application analog 2 out of 3 voter.

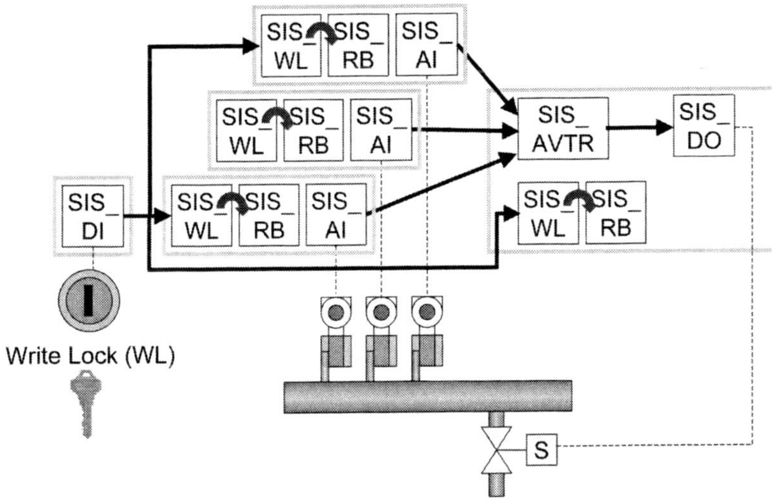

All devices will require a "Write Lock" block to allow them to be changed or altered as required. The Safety Discrete Input block (SIS-DI) serves the function of enabling and disabling the SIS-AI block in each of the three transmitters. Of course the Write Lock block has the necessary security features to insure that it is not inadvertently changed by someone in error. Since the Resource Block is the "switch" that turns a device on or off, it is activated or changed from OOS (Out Of Service) to AUTO (Automatic) or Cascade as required to turn the device on once any changes have been made. The SIS-AI blocks then record the process condition and communicate their results to a new SIS-AVTR (Analog voter) block that compares the readings against configured limits and when 2 of the three devices are outside those bounds activates the SIS-DO (discrete output) block to change the state of the final control element.

Implementation of Fieldbus safety systems will likely follow the same adoption cycle as the original H1 devices with devices being available some time before the associated host systems capable of taking advantage of their new functionality and features.

9.2 Electronic Device Description Language (EDDL)

The majority[1] of today's "smart" analog instruments use EDDL (Electronic Device Description Language), often shortened to DD (Device Description) in conversation. EDDL is the base technology underlying the HART, Profibus PA, and FOUNDATION™ Fieldbus protocols by defining a significant part of their User Layer.

The original EDDL specification as defined by IEC 61804 Part 2 contains little support for graphics; thus, the interface and degree of integration between field device and hosts was not only limited, but the level of integration and information that could be presented to a user varied between each installation depending on the host being used.

One technology that was developed to attempt to overcome these limitations was FDT/DTM (Field Device Tool/Device Type Manager). The FDT/DTM technology uses EDDL as the basis for its definitions so that device manufacturers can define the "look and feel" of how the user accesses the device information for other than the Process Variable (PV) and related signals, such as those used to configure and calibrate the device.

The HART Communications Foundation, Profibus International, and the Fieldbus Foundation also realized that users required improved interfacing with their devices and as a result have developed through a joint activity a number of graphical support functions to EDDL while maintaining 100% backwards compatibility with the original specifications prepared in the early 1990s. The key new capabilities of EDDL include the following:

- lets images be presented with installation/maintenance information

1. Many discrete devices also use EDDL; however, they are not yet in the majority.

- uses charts and menus to present and input data more consistently.
- uses graphs as X-Y plots to plot one variable against another such as a radar waveform (Figure 9-3) or valve signature (Figure 9-4).

Figure 9-3 – Radar level gauge.

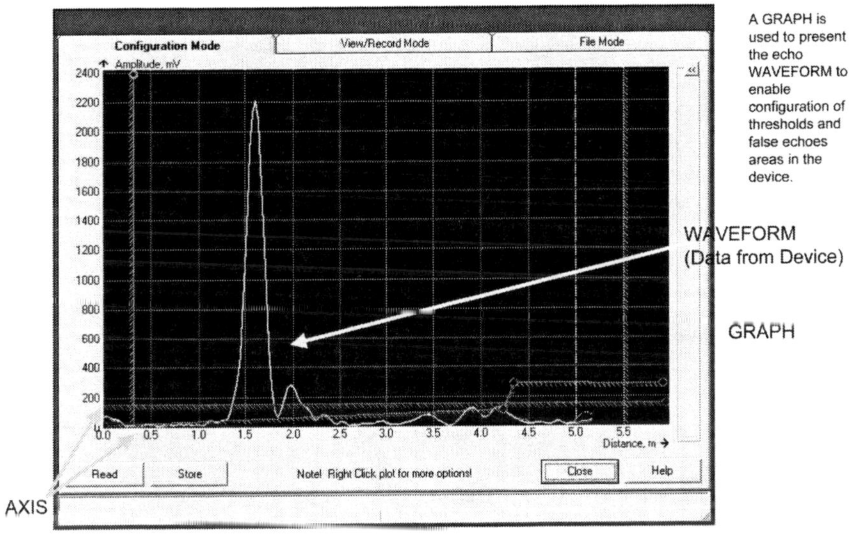

- provides basic mathematics for a first-order equation of the form y=mx+b as also seen in Figure 9-4
- offers history/archive to provide information over time

Other less obvious enhancements that will be "invisible" to end users have also been made to improve the ability for devices and host systems to communicate.

Figure 9-4 – Valve signature with best fit line.

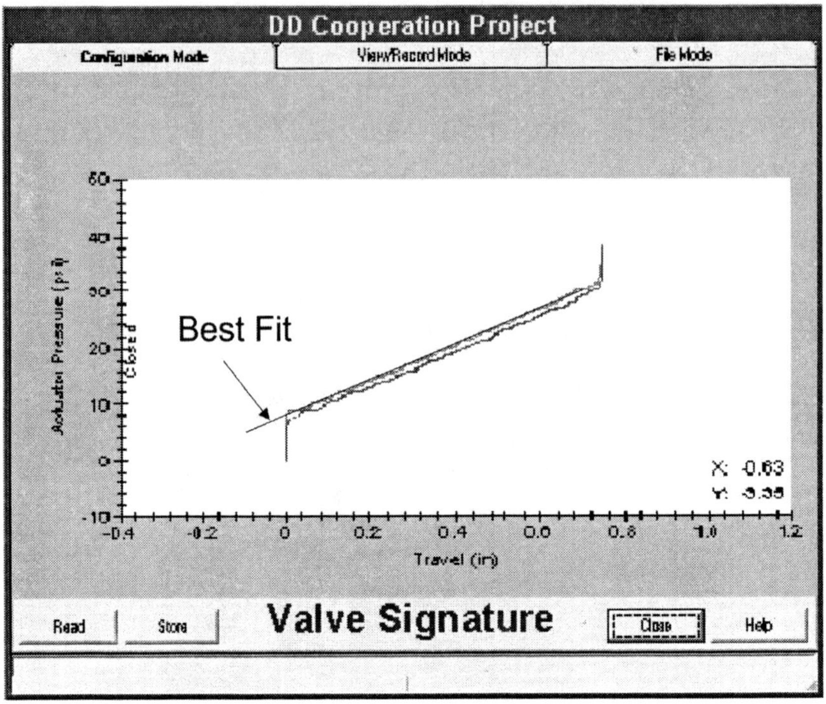

Encouraged by the success of this first project, the three organizations that rely on EDDL technology are now working with the OPC Foundation to add more features and capabilities to EDDL for extension beyond the traditional control system environment.

Appendix A – Nomenclature

	Description	Units
A	Attenuation	decibels (dB)
C_c	Cable capacitance	nanofarads (nF)
C_D	Capacitance of device	nF
C_{eq}	Equivalent capacitance	nF
C_i	Residual capacitance	nF
C_S	Capacitance of wire for network	nF
C_T	Total capacitance of the network	nF
E	Error in process loop (setpoint–PV)	Process units
f_r	Frequency range	hertz (Hz)
I_D	DC current draw of field device with largest minimum voltage	milliamps (mA)
I_i	Interconnection current	mA
I_k	Maximum network current	mA
I_N	Remaining or net current on network	mA
I_o	Power supply current	mA
K_p	Process gain	Dimensionless
L_c	Cable inductance	millihenries (mH)
L_i	Residual inductance	mH
L_{Maxx}	Maximum length of cable x	M
L_{Maxy}	Maximum length of cable y	M
L_S	Spur cable length	meters (m)
L_T	Trunk/home-run cable length	m
L_{TMAX}	Maximum voltage of trunk cable	volts (V)
L_x	Length of cable x	M
L_y	Length of cable y	M

N_C	Number of communications with the HMI	
N_P	Number of Publishers (devices on the network)	
P_i	Interconnection power	Watts (W)
P_o	Power supply power	W
R_c	Cable resistance	ohms/km (Ω/km)
R_S	Spur cable resistance	Ω/km
R_T	Trunk/home-run cable resistance	Ω/km
t_{LOAD}	Time to execute all Function Blocks in loop	Milliseconds (ms)
V_D	DC voltage available at the field device	V
V_i	Interconnection voltage	Volts
V_{Min}	Minimum voltage at device	V
V_{MIN}	Largest minimum voltage of all field devices	V
V_o	Power supply voltage	Volts
V_{PS}	Power supply voltage	V

Appendix B – Fieldbus Foundation Specification List

FF-103 Foundation Specification - Common File Format

FF-131 Foundation Specification - Standard Tables

FF-581 Foundation Specification - System Architecture

FF-586 Foundation Specification - HSE Presence

FF-588 Foundation Specification - Field Device Access (FDA) Agent

FF-589 Foundation Specification - HSE System Management

FF-593 Foundation Specification - High Speed Ethernet Redundancy

FF-801 Foundation Specification - Network Management

FF-803 Foundation Specification - HSE Network Management

FF-806 Foundation Specification - Data Link Protocol Specification Bridge Operation Addendum

FF-816 Foundation Specification - 31.25 kbits/sec Physical Layer Profile

FF-821 Foundation Specification - Data Link Services Subset

FF-822 Foundation Specification - Data Link Protocol Specification

FF-831 Fieldbus Power Supply Specification

FF-870 Foundation Specification - Fieldbus Message Specification

FF-875 Foundation Specification - Fieldbus Access Layer (Services and Protocol)

FF-880 Foundation Specification - System Management

FF-883 Foundation Specification - System Management Addendum for Software Download

FF-890 Foundation Specification - Function Block Application Process Part 1

FF-891 Foundation Specification - Function Block Application Process Part 2

FF-892 Foundation Specification - Function Block Application Process Part 3

FF-893 Foundation Specification - Function Block Application Process Part 4

FF-894 Foundation Specification - Function Block Application Process Part 5

FF-900 Foundation Specification - Device Description Language

FF-940 Foundation Specification - Communication Profile

FF-941 Foundation Specification - HSE Profile

Appendix C – Bibliography

C.1 Printed Materials

ANSI/ISA–50.02, Part 2-1992- *Fieldbus Standard for Use in Industrial Control Systems. Part 2: Physical Layer Specification and Service Definition*, ISA - The Instrumentation, Systems, and Automation Society, Research Triangle Park, NC, 1992.

ANSI/ISA-50.02, Part 3, 1997, *Fieldbus Standard for Use in Industrial Control Systems Part 3: Data Link Service Definition*, ISA - The Instrumentation, Systems, and Automation Society, Research Triangle Park, NC, 1997.

ANSI/ISA-50.02, Part 4, 1997, *Fieldbus Standard for Use in Industrial Control Systems Part 4: Data Link Protocol Specification*, ISA - The Instrumentation, Systems, and Automation Society, Research Triangle Park, NC, 1997.

ANSI/ISA-50.02, Part 5-1998, *Fieldbus Standard for Use in Industrial Control Systems Part 5: Application Layer Service Definition*, ISA - The Instrumentation, Systems, and Automation Society, Research Triangle Park, NC, 1998.

ANSI/ISA-50.02, Part 6-1998, *Fieldbus Standard for Use in Industrial Control Systems Part 6: Application Layer Protocol Specification*, ISA - The Instrumentation, Systems, and Automation Society, Research Triangle Park, NC, 1998.

Berge, J., *Fieldbuses for Process Control: Engineering, Operation, and Maintenance*, ISA - The Instrumentation, Systems, and Automation Society, Research Triangle Park, NC, 2001.

Fieldbus Application Guidelines for the Process Industry, Engineering Equipment and Material Users Association, London, UK, 1997.

Fieldbus Book, A Tutorial, Technical Information bulletin, TI 38K02A01-01E, Yokogawa Electric Corporation, Tokyo, Japan, May 2001.

Fieldbus Preliminary Application Note on Intrinsic Safety, Revision 1.1, Fieldbus Foundation, Austin, TX, 1995.

Fieldbus Wiring Design and Installation Guide (Available at www.relcom-inc.com), Relcom Inc., Forest Grove, OR.

Foundation Fieldbus Discovery Course, Southern Alberta Institute of Technology, Calgary, Alberta, 2001.

Goeldner, H.D., U. Johansmeyer, F. Schebstadt, and H. Storck, *PTB W39* and *53a*, Wirtschaftsverlag NW, Bremerhaven, Germany, 1989.

IEC-61158–2, *Fieldbus Standard for Use in Industrial Control Systems. Part 2: Physical Layer Specification and Service Definition*, International Engineering Consortium, Chicago, IL, 1993.

IEC 65C/178/CDU–IEC 61158–3, *Data Link Layer–DLL Service. Part 3*, International Engineering Consortium, Chicago, IL, 1999.

IEC 65C/179/CDU–IEC 61158–4, *Data Link Layer–DLL Protocol. Part 4*, International Engineering Consortium, Chicago, IL, 1999.

ISA-TR50.02, Part 9-2000, *Fieldbus Standard for Use in Industrial Control Systems: User Technical Report*, ISA - The Instrumentation, Systems, and Automation Society, Research Triangle Park, NC, 2000.

AN9027 *FNICO Non-Incendive Fieldbus System*, MTL Instruments Application Note, June 2004.

Fieldbus Foundation Application Guides

AG-140 *Wiring and Installation 31.25 kbit/s, Voltage Mode, Wire Medium Application Guide*, Fieldbus Foundation, Austin Texas, 1996

AG-163 *31.25kbit/s Intrinsically Safe Systems Application Guide*, Fieldbus Foundation, Austin Texas, 1996

AG-181 *FOUNDATION™ Fieldbus System Engineering Guidelines*, Fieldbus Foundation, Austin Texas, 2004

C.2 Web Sites

www.fieldbus.org
www.isa.org/fieldbus
www.iceweb.com.au/fieldbus
www.wib.nl
http://ourworld.cs.com/rahulsebos

C.3 Fieldbus Certified Training Centers

Waseda University	Kitakyushu, Japan
Shipping & Transport College, Brielle	Brielle, The Netherlands
Lee College	Baytown, Texas
Southern Alberta Institute of Technology	Calgary, Alberta
Tri-State University	Angola, Indiana
Singapore Polytechnic	Singapore
SINOPEC Yanshan Simulation Center (YSC)	Beijing, China

C.4 Free Segment Design Tools

Excel Spreadsheet, "FFSegCheck" – Yokogawa Corporation
http://www.yokogawa.com/fbs/fbs-download-en.htm

Excel Spreadsheets, Hawke Systems –
http://www.ehawke.com/fieldbus/fieldbus_download.htm

Application, Pepperl+Fuchs, http://www.segmentchecker.com

Application, Emerson Process Management "Segment Checker,"
www.emersonprocess.com/systems/support/segment

Appendix D – Acronyms

D.1 Acronyms

Acronym	Description
non-FF Acronyms in Book	
AWG	American Wire Gauge
DCS	Distributed Control System
DCS	Digital Control System
HMI	Human Machine Interface
I/O	Input/Output
IEC	International Electrotechnical Commission
IS	Intrinsically Safe
ISA	Instrumentation, Systems, & Automation Society
ISO	International Standards Organization
mA	milliamps
NIS	Non-Intrinsically Safe
OLE	Object Linking and Embedded
OPC	OLE for Process Control
PTB	Physikalisch Technische Bundesanstalt
SCADA	Supervisory Control And Data Acquisition
SP	Set Point
VDC	Volts Direct Current
XML	eXtensible Mark-up Language

Acronym	Description
FF Acronyms in Book	
A	Arithmetic
AA	Analog Alarm
AHI	Analog Human Interface
AI	Analog Input
AO	Analog Output
Auto	Automatic
BG	Bias/Gain Station
C	Calculate
CAO	Complex Analog Output
Cas	Cascade
CDO	Complex Discrete Output
CS	Control Selector
DA	Digital Alarm
DC	Device Control
DD	Device Description
DHI	Digital Human Interface
DI	Discrete Input
DL	Data Link
DO	Discrete Output
DT	Deadtime
FFB	Flexible Function Block
FISCO	Fieldbus Instrinsically Safe Concept
H1	Fieldbus Low Speed Data Network
HSE	High Speed Ethernet
Iman	Initialization Manual
IP	Internet Protocol
IS	Input Selector
IT	Integrator (Totalizer)

Acronym	Description
LAS	Link Active Schedule
LL	Lead Lag Controller
LM	Link Master
LO	Local Override
LS	Link Scheduling
MAC	Management Access Control
MAI	Multiple Analog Input Block
Man	Manual
MAO	Multiple Analog Output Block
MDI	Multiple Discrete Input Block
MDO	Multiple Discrete Output Block
MIB	Management Information Base
ML	Manual Loader
OOS	Out Of Service
OS	Output Splitter Block
PD	Proportional Derivative
PD	Process Data
PI	Pulse Input
PID	Proportional Integral Derivative
RA	Ratio
RB	Resource Block
RCas	Remote Cascade
ROut	Remote Output
RX	Receive/Receiver
SC	Signal Characterizer
SOPID	Step Output PID
SRG	Set Point Ramp Generator
SS	Signal Splitter
STP	Shielded Twisted Pair(s)

Acronym	Description
T	Timer
TCP	Transmission Control Protocol
TD	Transducer Block
TX	Transmit/ Transmitter
UDP	User Datagram Protocol
UTP	Unshielded Twisted Pair(s)
VCR	Virtual Communication Resource
VFD	Virtual Field Device
XD	Transducer

D.2 Foundation Fieldbus Acronyms

Acronym	Description
Ack	Acknowledge
AE	Application Entity
AI	Analog Input
AO	Analog Output
AP	Application Program
AP	Application Process
APDU	Application Layer Protocol Data Unit
AR	Application Relationship
AREP	Application Relationship End Point
ARP	Ethernet Address Resolution Protocol
ARPM	Application Relationship Protocol Machine
ASN.1	Abstract Syntax Notation 1
BOOTP	Bootstrap Protocol
CAP	Control Application Process
Cas	Cascade

Acronym	Description
CD	Compel Data
CFF	Capability File Format
CFF	Common File Format
CNF	Confirmation
CNLS	Connectionless
COM	Communication
Config	Configuration
CONN	Connection-Oriented
CSMA/CD	Carrier Sense Multiple Access / Collision Detect
DAP	Device Application Process
DCS	Distributed Control System
DD	Device Description
DDL	Device Description Language
DevId	Device Identifier
DHCP	Dynamic Host Configuration Protocol
DI	Discrete Input
DL	Data Link
DLCEP	Data Link Connection End Point
DLL	Data Link Layer
Dlme	Data Link Layer Management Entity
DLPDU	Data Link Protocol Data Unit
DLSAP	Data Link Service Access Point
DMA	Direct Memory Access
DNS	Do Not Select
DO	Discrete Output
DPS	Draft Preliminary Specification
DS	Data Structure
DTC	Data Transfer Confirmed
DTU	Data Transfer Unconfirmed

Acronym	Description
DUT	Device Under Test
DV	Dynamic List of Variable List
EFD	(High Speed) Ethernet Field Device
EGP	Exterior Gateway Protocol
EP	Ethernet Presence
EPI	Ethernet Presence Interface
ERR	Error
EU	Engineering Unit
FAS	Fieldbus Access Sublayer
FB	Function Block
FBAP	Function Block Application Process
FDA	Field Device Access
FDA	Field Device Application (Agent)
FDA	Field Device Access Agent
FDA	Field Device Access
FDL	Fieldbus Data Link Layer
FF	Fieldbus Foundation
FFB	Flexible Function Block
FMS	Fieldbus Message Specification
FMS	Fieldbus Messaging System
FS	Final Specification
FSA	Fail-Safe Active
FSA	Fault Statefail -Safe Active
FSPM	FDA Service Protocol Machine
Gen	Generic
H1	Hunk 1
HFD	HSE Field Device
HL	Link Designator of a four-octet LONG DL-address
HL.N.S	Four-octet LONG DL-address

Acronym	Description
HLD	Higher Level Device
HMA	HSE Management Agent
HSE	High Speed Ethernet
I/O	Input / Output
IA	Initialization Acknowledge
IANA	Internet Assigned Numbers Authority
ICMP	Internet Control Message Protocol
ICS	Implementation Conformance Statement
ID	Identifier
Id	Identifier
IEEE	Institute of Electrical and Electronics Engineers
IETF	Internet Engineering Task Force
IGMP	Internet Group Management Protocol
IMan	Initialization Manual
IND	Indication
INFO	Informational
IP	Internet Protocol
IPv4	"Internet Protocol, Version 4"
IPv6	"Internet Protocol, Version 6"
IR	Initialization Request
IT	Information Technology
IUT	Implementation Under Test
LAN	Local Area Network
LAS	Link Active Scheduler
LD	Linking Device
LDM	Last Delivered Message
LLC	Logical Link Control
LME	Layer Management Entity
LO	Local Override

Acronym	Description
LOP	Local Operator Panel
LRE	LAN Redundancy Entity
LS	Link Schedule
LSB	Least Significant Bit
LUV	Last Usable Value
MAC	Medium Access Layer
MAC	Medium/Media Access Control
MAI	Multiple Analog Input
MAO	Multiple Analog Output
MAU	Medium Attachment Unit
Max.	Maximum
Mb/s	Megabits per second
MDI	Multiple Digital Input
MDI	Multiple Discrete Input
MDO	Multiple Digital Output
MDO	Multiple Discrete Output
MDS	Medium Dependent Sublayer
Mfr.	Manufacturer
MIB	Management Information Base
MIB-II	Management Information Base II
MSB	Most Significant Bit
Msg	Message
MVC	Multi-Variable Container object
N.S.	Sub-link Selector of a four-octet LONG DL-address
NI	Not Invited
NM	Network Management
NMA	Network Management Agent
NMIB	Network Management Information Base
NST	Network Status Table

Acronym	Description
NTP	Network Time Protocol
OD	Object Description
OD	Object Dictionary
OSI	Open System Interconnection
PD	Physical Device
PD Tag	Physical Device Tag
PDU	Protocol Data Unit
PhPDU	Physical Layer Protocol Data Unit
PHY	Physical Layer
PICS	Protocol Implementation Conformance Statement
PID	Proportional Integral Derivative (control algorithm)
PLC	Programmable Logic Controller
PN	Probe Node PDU
PR	Probe Response PDU
PROP	Proposed
PS	Preliminary Specification
PT	Pass Token
RCas	Remote Cascade
RED	Redundant
Ref	Reference
REQ	Request
RFC	Request For Comments
RFU	Reserved for Future Use
ROut	Remote Output
RSE	Remote Single Layer Embedded
RSP	Response
RT	Return Token
SM	System Management
SMI	Structure and Identification of Management Information

Acronym	Description
SMIB	System Management Information Base
SMK	System Management Kernel
SMKP	System Management Kernel Protocol
SMKPM	System Management Kernel Protocol Machine
SMPM	Socket Mapping Protocol Machine
SNMP	System Network Management Protocol
SNTP	Simple Network Time Protocol
STD	Standard
SUT	System Under Test
TCP	Transmission Control Protocol
TCP/IP	Transmission Control Protocol / Internet Protocol
TD	Time Distribution
TOS	Type Of Service (IP definition)
TTCN	Tree & Tabular Combined Notation
TTL	Time To Live (IP definition)
UDP	User Datagram Protocol
VCR	Virtual Communication Relationship
VCRL	Virtual Communication Relationship List
VFD	Virtual Field Device

Index